THE AI SUCCESS BLUEPRINT

WHY 80% FAIL AND HOW YOUR BUSINESS CAN WIN

DARREN BEKKER

PREFACE

Every business wants AI now. Most businesses do not know why, exactly. They know competitors are doing something. They have read the McKinsey, Gartner reports. They have seen the headlines about efficiency gains and competitive advantage. So, they do what organizations do when everyone else is moving: they move too, hoping direction will become clear along the way.

It does not.

Over a decade of transformation work, across digital initiatives and agile adoptions and operational overhauls, taught me one thing: the fundamentals never change. It doesn't matter if it is AI or any other technology. Success comes down to three things working together. Clear goals that people understand. Humans who believe the change makes sense. Disciplined execution focused on outcomes that matter.

That is what this book is about. Not the technology itself, though we will cover that. The process. The approach. The thinking that separates the 20% who succeed from the 80% who pour resources into initiatives that deliver little beyond expensive lessons.

The statistics bothered me enough to research the claims and author this book. Eighty percent. That is not a rounding error. That's a systemic problem. Behind each failed initiative are real people who tried. Executives who believed they were making the right call. Teams who worked late building something that ultimately didn't matter. Resources that could have gone to efforts with actual impact.

They didn't fail because they lacked intelligence, budget, or commitment. They failed because nobody gave them a clear path through complexity, because the process was unclear. AI got positioned as a technology problem when it's a business transformation problem. Companies started with tools instead of strategy, with pilots instead of purpose.

This book simplifies that process. It will not teach you to build neural networks or explain transformer architecture. Other books do that better. This gives you something more useful: a blueprint. A clear, practical approach to AI that works regardless of your industry, geography, or company size.

You will learn why most initiatives fail and how to avoid those traps. How to build strategy that connects AI to measurable business outcomes. How to prepare your data and your people. How to choose partners and tools that fit your actual needs rather than someone's sales targets. How to move from pilot to scale without losing focus. How to measure real ROI instead of vanity metrics. How to sustain success through governance and culture.

I have seen what happens when organizations get this right. Productivity improves, yes. Customer experiences transform. Those matter. What matters more is teams rediscovering the excitement of solving meaningful problems. People feel like technology serves them rather than threatening them. Organizations building capability that compounds rather than implementing point solutions that age out.

AI is not just another technology wave. It is fundamentally changing how work gets done, how decisions get made, how value gets created. The leaders who thrive will not necessarily know the most about machine learning algorithms. They will be the ones who can take complexity and make it simple. Who can connect technology to business value? Who can build organizations that adapt as technology evolves?

Whether you are a CEO wondering where to start, a transformation leader tired of initiatives that stall, or a middle manager trying to make sense of competing priorities, this book is for you. For anyone exhausted by buzzwords and ready for results. For the 80% who have struggled and the

20% who want to stay ahead. For those who believe success comes not from building complexity but from building clarity.

If you have felt overwhelmed by AI, uncertain where to focus, unsure how to separate signal from noise, this book was written for you.

Let us cut through the confusion. Let us focus on what works. Let us build something that actually matters.

Darren Bekker
Cape Town, South Africa

TABLE OF CONTENTS

INTRODUCTION

WHY 80% FAIL (AND WHY YOU WON'T)

The problem is not that AI does not work. The problem is how organizations approach it.

Most start in the wrong place. They chase technology before strategy, on tools before problems. They talk about models and algorithms before they talk about people and processes. They confuse activity with progress, pilot projects with transformation, data science teams with business outcomes.

Somewhere between the hype and the implementation plan, purpose gets lost. The "why" becomes fuzzy. The connection between investment and value becomes theoretical. Teams build impressive capabilities that do not solve actual problems. Executives approve budgets without clear success criteria. Everyone stays busy, but nothing really changes where it matters.

The result? Eighty percent of AI transformations do not deliver meaningful business impact.

Not because technology is flawed. AI works, demonstrably, across countless applications. Technology is not the bottleneck. The organizational approach is. Companies forget why they are doing this in the first place. They get seduced by possibility and lose sight of practicality. They improve for sophistication instead of value.

This book exists to change that pattern.

The AI Success Blueprint isn't a technical manual. You will not learn to code machine learning models or architect data pipelines. Other resources cover that ground thoroughly. This is something different. A field guide for leaders who need to turn AI from abstract concept into competitive advantage. A practical roadmap through complexity.

If you have wondered where to start, which initiatives matter, how to build a business case that withstands scrutiny, how to prove ROI, how to manage organizational resistance, this book is your compass through that territory.

Here is what you'll know at the end: why most projects fail and how to avoid those traps. How to build strategy that connects to business goals. How to choose use cases worth pursuing versus ones that sound impressive but deliver little. How to prepare your organization, your data, your people. How to scale what works without getting stuck in pilot purgatory. How to measure value that matters rather than metrics that look good in presentations.

This isn't about "doing AI." Plenty of companies are doing AI. They are running pilots, hiring data scientists, buying platforms, and attending conferences. Activity isn't the problem. Impact is. This book is about winning with AI. Creating transformation that improves how your business works, competes, and grows.

The industry has a secret most won't tell you directly: it profits more from confusion than clarity. Every week brings another breakthrough tool, another model that promises revolution, another platform that claims to solve everything. Without strategy, these tools are like race cars without drivers. Powerful and sophisticated but useless.

The companies succeeding with AI do not necessarily have bigger budgets. They have better processes. They focus on problems worth solving rather than solutions looking for problems. They enable people rather than threatening them with replacement. They measure outcomes worth tracking rather than outputs that are easy to count. They understand AI is fundamentally about transformation, which means it is fundamentally about people, which means technology is the easy part.

Here is my philosophy, distilled from a decade watching organizations succeed and fail: AI success isn't about doing everything. It is about doing the right things consistently.

The organizations winning the AI race do not chase hype. They follow principles. They balance strategy with execution, technology with people, ambition with pragmatism. They move deliberately, learn continuously, scale thoughtfully.

That is what this book gives you. A structured approach refined from watching what works across industries, geographies, company sizes. Frameworks used by enterprises and startups alike, distilled into steps any organization can apply. Methods for building clarity where there's noise, creating adoption where there's fear, driving results where others stall.

Whether you are a CEO mapping strategy, a CIO managing implementation, or a team leader trying to navigate competing priorities, you will find guidance here. Not theory for theory's sake. Practical steps grounded in reality. Tools you can use at once. Frameworks you can adapt to your context.

We are living through one of those rare moments when technology fundamentally reshapes how work gets done. Nothing this transformative has happened since the internet changed the world. These inflection points create enormous opportunities for those who understand them and enormous risk for those who do not.

The organizations that get AI right will build extraordinarily competitive advantages. Those that hesitate or follow the wrong approach risk irrelevance. Not at once. Not dramatically. Just a slow erosion of position as competitors becomes more efficient, more responsive, more capable of delivering value.

AI will not replace you. Leaders who understand AI will.

That is why you're here. To join the 20% who succeed. To build approaches that work. To create value for your people, your customers, your organization. To turn complexity into clarity, confusion into confidence, investment into impact.

Most of the research referenced in this book comes from reputable sources: McKinsey, Gartner, MIT, Stanford, Harvard Business Review. When you see statistics about failure rates or case studies about implementation approaches, they are grounded in actual research. Some examples are synthesized to illustrate points without finding specific companies. The patterns are real; the details are sometimes disguised to protect confidentiality and broaden applicability.

If you want to go deeper, join the community at theaisuccessblueprint.com. Free resources. All the tools referenced in this book. A place to ask questions, share wins, discuss challenges with others on the same journey. Transformation is easier when you are not doing it alone.

Let's begin.

CHAPTER 1

THE TRILLION DOLLAR GRAVEYARD

The VP of Operations looked tired.

"We had everything," she said, "Executive buy-in. A £4.2 million budget. The Best Consultants. The best AI platform money could buy." She paused, then said. "We had everything except the one thing that mattered."

"We didn't have a clue what problem we were actually trying to solve."

This manufacturing company, a 200-person operation making precision components for the automotive industry, had done what countless organizations did when the AI revolution knocked on their door. They'd panicked. They had watched competitors announce AI initiatives and read breathless articles about machine learning transforming supply chains. The team sat through vendor presentations promising 40% efficiency gains. And then, in a moment of collective corporate panic, they had signed a large, high value contract.

The plan had sounded reasonable enough. Deploy AI across the production floor to predict equipment failures, optimize inventory, and reduce waste. The kind of transformation that looks brilliant in a PowerPoint deck. The consultants had been enthusiastic, naturally. The software vendor had case studies. The board had been impressed.

Eighteen months later, they had exactly three things to show for their investment: a massive Cloud bill they did not quite understand, a custom dashboard that nobody looked at, and a team that had learned to hate the phrase "Artificial Intelligence."

"The models were probably working," the VP said, and there was something almost wistful in her voice. "I mean, they were spitting out predictions, numbers and even recommendations. But here is the thing, nobody trusted them. The plant managers, guys who have been running those machines for twenty years, would look at the AI's maintenance predictions and ignore them. They had their own way of knowing when a machine was about to fail. And honestly? Their instincts were better."

This was not a story about technology failing. The technology had worked exactly as designed. This was a story about an organization that had bought a solution before understanding its problem, which had confused spending money with creating value, which had somehow believed expensive software could compensate for the hard work of actually changing how people work.

This is where I should tell you that this story is unusual. An outlier. The exception that proves the rule, but I cannot.

The details change, sometimes it's customer service AI that doesn't understand customers, sometimes it's predictive analytics that nobody uses to predict anything, sometimes it's a chatbot that makes customers angrier than the problem it was meant to solve, but the plot remains depressingly consistent.

The numbers do not lie.

Let's talk about the elephant in the room, the truth that everyone in the AI industry knows but nobody particularly wants to say aloud: most AI transformations fail.

According to recent research, companies spent over $350 billion on AI initiatives in 2025 alone. By 2030, cumulative global investment in AI is projected to exceed $2 trillion. Yet study after study reveals the same uncomfortable truth: somewhere between 70% and 85% of AI projects fail to deliver meaningful business value.

Think about that for a moment. We are creating the most expensive graveyard in business history, one abandoned chatbot, one dormant machine learning model, and one "strategic AI initiative" at a time.

Here's what makes this particularly maddening: AI works. Technology itself is not the problem. Recommendation engines can predict what customers want. Computer vision systems genuinely can detect defects faster than human inspectors. Natural language processing honestly does understand context better than keyword search. The algorithms are fine. Math is sound. The technology is proven.

So why does it keep failing?

THE SUCCESS PARADOX

There is something genuinely strange happening in the AI transformation space, a paradox that should concern anyone thinking about implementing AI in their organization. The same technology that powers Google's search, Amazon's recommendations, and Netflix's content suggestions somehow becomes useless when it crosses the threshold into most ordinary businesses.

This is not a problem with capabilities. A mid-sized retailer implementing product recommendations is not trying to do anything more complex than what Amazon does. A regional bank using chatbots is not attempting anything more sophisticated than what is already working at major financial institutions. Technology is the same. The problems are similar. The budgets are often adequate.

Yet Amazon's recommendation engine generates 35% of their revenue, while your company's recommendation system might generate complaints from the IT team about server costs.

This is the AI Success Paradox: proven technology, implemented by capable people, with adequate resources, consistently does not deliver value. It is as if we're all trying to bake bread using the same recipe that works perfectly in professional bakeries, but somehow, our loaves keep coming out flat.

The paradox deepens when you look at who is failing. This is not about small companies lacking resources or expertise. Some of the most spectacular AI failures have happened at organizations with unlimited budgets and access to the world's best talent. A major US healthcare system spent three years and tens of millions of dollars on an AI diagnostic system that doctors simply refused to use. A European bank's customer service AI became so notorious for unhelpful responses that customers started requesting human agents before even trying the bot. An Asian manufacturer's predictive maintenance system was so unreliable that maintenance teams developed an unofficial policy: check the equipment themselves, ignore the AI.

These were not underfunded projects. They were major transformation initiatives with executive sponsorship, resolute teams, and consulting support. They had everything except the result they wanted.

The conventional wisdom about why AI projects fail tends to focus on technical issues. Not enough data. Poor data quality. Wrong algorithms. Inadequate computing power. And yes, these things matter. But they're not the main story. If technical problems were the primary obstacle, we would expect AI success rates to improve as the technology matures and becomes more accessible. Instead, failure rates have remained stubbornly consistent even as the tools have become dramatically better and cheaper.

A research team at Stanford spent two years studying AI implementations across 300 companies. What they found was illuminating and slightly uncomfortable. The technical execution of most failed AI projects was actually rather good. The models worked, the infrastructure held up and the predictions were accurate.

The projects failed anyway.

The organizations implementing them had not done the hard, unglamorous work that makes AI valuable: clearly defining business problems, preparing their data, training their people, managing the change, measuring the right outcomes. They had confused buying technology with creating transformation.

WHAT THIS BOOK PROMISES, AND WHAT IT DOESN'T

I am not going to tell you that AI transformation is easy. It is not. I am not going to promise that following this blueprint guarantees success. Nothing guarantees success. I am not going to claim that AI is the answer to every business problem, because it is not.

What I can promise is honesty about what works and what does not, based on over a decade of watching organizations transform themselves. I have seen enough failures to recognize the patterns that lead to them. I have seen enough successes to understand what separates them from the graveyard projects.

This book exists because of frustration. Most AI books and consultants either oversimplify or overcomplicate. They either tell you AI is magic that will solve everything, or they drown you in technical details that matter only to data scientists. Reality lives somewhere in between. AI is powerful but not magical. It is technical but comprehensible. It requires rigor but not a PhD.

The successful 20% of AI transformations share common patterns. They start differently. They think differently. They measure differently. They fail differently too, but they fail in ways that generate learning rather than just burning money.

This book is structured around a simple premise: if 80% of AI transformations fail for predictable reasons, then understanding and avoiding those reasons puts you in the successful minority. You do not need to be smarter than everyone else. You do not need more money than everyone else. You need to avoid making the same mistakes that everyone else makes.

Part I of this book examines the failure epidemic in detail. Not to be depressing, but because you cannot avoid pitfalls you do not recognize. We will look at the seven deadly sins of AI transformation, the cognitive biases that sabotage good intentions, and the early warning signs that distinguish doomed projects from successful ones.

Part II provides the blueprint itself: a maturity model to assess where you actually are, a strategic framework for deciding where to go, an implementation roadmap for getting there, and a change management approach that treats humans as the key to success rather than an obstacle to overcome.

Part III makes it practical. How to build your team. How to get your data ready. How to measure success in ways that matter. How to create an organization that does not just implement AI projects but builds lasting capability.

Throughout, I will share real examples and situations from real organizations. Manufacturing firms and healthcare systems and retailers and banks and nonprofits and government agencies. Small companies with ten employees and large enterprises with ten thousand. Different industries, different continents, different problems but remarkably similar patterns of failure and success.

WHY YOU SHOULD TRUST THIS BOOK (AND WHY YOU SHOULDN'T)

I have spent a decade managing and leading organizational transformations. I have reviewed numerous MIT, Gartner, and McKinsey reports. I have also studied failures and successes. There are common patterns in both successes and failures.

My only interest is in sharing what works, based on what I have observed, so that fewer organizations end up staring at the wreckage of their good intentions.

You should not trust this book blindly. You shouldn't trust any book blindly. AI transformation is contextual. What works brilliantly for an organization or company might flop at a hospital. What succeeds in Germany might fail in Japan, not because the AI is different but because the organizational culture is different. Your situation is unique in ways that matter.

So, here is what the book promises: I will give you frameworks, not formulas. Principles, not prescriptions. Tools for thinking, not step-by-step instructions that pretend your situation is identical to every other situation. I will tell you what the successful 20% consistently do differently, and the patterns observed. You will need to figure out how to adapt it to your specific context.

The successful 20% are not better at AI, they do not necessarily have bigger budgets. They are better at knowing where AI can help their organization and what is required to transform.

The rest of this book is about learning what they know.

Let us begin with the most important question: Why do smart organizations make such consistently bad decisions about AI? Until you understand the psychology of failure, no amount of technical knowledge will save you.

CHAPTER 2

THE SEVEN DEADLY SINS OF AI TRANSFORMATION

The presentation looked perfect.

Forty-three slides, each one crisper than the last. Market analysis showing competitors' AI investments, technical architecture diagrams with impressive complexity. Financial projections showing 300% ROI within eighteen months and an implementation timeline spanning twelve months with clearly defined phases. It also included risk mitigation strategies for every conceivable obstacle.

The Chief Strategy Officer had spent three months preparing this, working with a Big Four consulting firm that specialized in AI transformation. The recommendation was clear: invest $18 million in an AI-powered customer intelligence platform that would revolutionize how the company understood and served its market.

The board approved it unanimously.

Two years later the project had been quietly wound down six months earlier after delivering almost nothing of value. So, what happened?

They committed every sin in the book. They just did not know there was a book.

There is a book. It is written in the wreckage of failed implementations across every industry and geography. It keeps telling the same story, describing the same patterns, and documenting the same mistakes. The seven deadly sins of AI transformation are not obscure or subtle. They are obvious in retrospect, even predictable. Yet organizations keep committing them with remarkable consistency.

Understanding these sins will not guarantee success, but it will help you avoid the most common paths to failure. Think of them as the clear markers on the trail, the signs warning you that ahead lies a cliff edge.

SIN 1: STARTING WITH TECHNOLOGY INSTEAD OF BUSINESS PROBLEMS

Here is how it usually begins.

Someone reads about AI in the Harvard Business Review or hears about it at a conference. Technology sounds transformative. Neural networks, machine learning, natural language processing, terms that carry weight and promise. A meeting gets scheduled. "We should explore AI opportunities." Everyone nods. A committee forms.

The committee starts with technology. What can AI do? What are the latest capabilities? What are our competitors implementing? They invite vendors to present. Each presentation shows impressive capabilities and compelling use cases. The committee gets excited.

Then someone asks the fatal question: "Where should we apply this in our business?"

The sequence is backwards. They have started with a solution and are now searching for problems they might solve. This is technology-first thinking, and it dooms more AI initiatives than perhaps any other single factor.

A mid-sized logistics company fell into this trap completely. They decided AI was strategically important and allocated $8 million to "develop AI capabilities." They hired data scientists. They invested in infrastructure. They ran pilot projects in four different areas: route optimization, demand forecasting, warehouse automation, and customer service.

None of the pilots addressed their most pressing business problems. Their biggest challenges were driver retention, aging vehicle fleet, and margin pressure from larger competitors. The AI pilots were intellectually interesting but operationally marginal. Two years later, they had impressive technical capability and negligible business impact.

The problem with starting with technology is that it creates solutions to search for problems rather than solutions addressing problems. And solutions to search for problems rarely find the right ones because search is constrained by what technology does well, not by what the business actually needs.

Business-first thinking inverts this. It starts with the question: What are our most significant business problems, the ones that meaningfully constrain growth or profitability? Then asks: Could AI potentially address any of these problems better than alternative approaches?

This sequence matters enormously. Starting with business problems, grounds the conversation in reality. It forces specificity about what success looks like. It creates natural criteria for evaluating whether AI is the right approach.

A pharmaceutical distribution company took this approach. They identified their core problem: inventory carrying costs were too high, but stock-outs were damaging customer relationships. They explored multiple potential solutions, including better forecasting processes, different supplier arrangements, and yes, AI-powered demand prediction.

They chose AI only after determining that their demand patterns were too complex for traditional statistical methods but regular enough that machine learning could improve predictions. Technology was selected because it fits the problem, not because it was fashionable.

The implementation succeeded. It delivered measurable improvement in inventory turnovers while reducing stock-outs. It worked because it was solving a real problem where AI was genuinely the best available approach.

Compare that to a hospitality company that implemented AI chatbots for customer service because "everyone's doing AI." They had not identified customer service as a problem area. Satisfaction scores were quite good. AI felt innovative, and the vendor pitch was compelling.

The chatbots frustrated customers who preferred human interaction for anything complex. Employees resented the implication that they could be replaced by algorithms. The technology worked fine; it just solved a problem the company did not actually have.

Technology-first thinking feels strategic because it is forward-looking, because it engages with innovation, because it sounds impressive in board presentations. "We're investing in AI capabilities" carries more weight than "We're improving inventory management." But strategic-sounding initiatives that do not solve real problems are just expensive distractions.

The test is simple: can you articulate the specific business problem you are solving in plain language without mentioning AI? If you cannot, if the problem statement requires AI terminology to make sense, you're probably starting from the wrong place.

SIN 2: BELIEVING VENDOR HYPE WITHOUT VALIDATION

Vendor demonstrations are designed to impress, and they do.

You watch as the AI processes thousands of data points in seconds, generating insights that would take analysts weeks. You see customer service queries handled seamlessly, fraud detected instantly, patterns identified that humans would never spot. The demos are slick, the technology is real, the capabilities are genuine.

What you are not seeing is how often the demo uses carefully curated data, optimized scenarios, and edge cases that have been smoothed away. You do not see the months of tuning required to get those results. You are not seeing what happens when the data is messy, the scenarios are ambiguous, or the context shifts from controlled demonstration to operational reality.

This is not deception. It is marketing. Vendors show you what is possible under ideal conditions. They are not obligated to dwell on what is typical under your conditions.

A regional bank learned this painfully. They watched vendor demonstrations of AI-powered loan underwriting that processed applications in minutes with accuracy rivaling experienced underwriters. The demos were impressive enough that they signed a $6 million contract without extensive validation.

Implementation revealed that the demo data had been clean, complete, and structured. The bank's actual data was none of those things. Applications often had missing information, inconsistent formatting, and ambiguous documentation. The AI that performed beautifully in demos struggled in production, regularly producing recommendations that underwriters had to override.

By the time they had invested another $3 million in adapting the system to their reality, they had something that worked marginally better than their existing process at higher cost. The vendor had not lied, they had just shown best-case performance and let the bank assume it was typical.

The antidote to vendor hype is independent validation. Before committing significant resources, evaluate the technology against your actual data, your real scenarios, your specific context. Not demos, not pilot data, but production reality.

This requires discipline because it slows the process down. Vendors will suggest that extensive testing is not necessary, that other clients have validated the approach, that delay means missing the opportunity. Organizations eager to move forward will rationalize why testing can be abbreviated.

But skipping validation to move faster is like skipping the foundation to build the house quicker. The speed gained early gets paid back with interest later.

A manufacturing company considering AI quality inspection insisted on a proof-of-concept using six months of their actual production data, including edge cases, anomalies, and the messy reality of their operation. Two of the three vendors they assessed performed far worse on real data than their demos suggested. The third vendor's technology held up, and that is who they contracted with.

The implementation succeeded in large part because they knew from validated testing that technology could actually manage their reality. No surprises, no disappointment, no expensive discovery that the demo performance did not translate.

Validation should answer specific questions:

Does this technology work with our data quality?
Can it manage our volume and complexity?
Does it perform acceptably in edge cases?
How much tuning and maintenance does it require?
What happens when something unexpected occurs?

These questions cannot be answered by vendor presentations or reference calls. They require firsthand testing with your actual situation.

The cost of thorough validation might be 5% of the overall project budget. The cost of discovering after implementation that the vendor's promised performance does not materialize in your environment can be the entire project budget plus organizational credibility.

Yet organizations routinely skip adequate validation, convinced by compelling demonstrations and confident vendor assurances. They sign contracts based on what they have been shown rather than what they have verified. They pay for those mistakes throughout implementation.

SIN 3: IGNORING THE DATA FOUNDATION REQUIREMENTS

AI runs on data the way cars run on fuel. Not just any data, but clean, structured, accessible, sufficient data. This seems obvious, yet organizations consistently underestimate what "good enough" data looks like.

The typical pattern: someone proposes an AI initiative. The question gets asked: "Do we have the data for this?" Someone from IT confirms that yes, the data exists. It is in various systems, it has been collected for years, there is plenty of it. The organization proceeds.

What they didn't ask: Is the data clean? Is it consistent across systems? Is it labeled correctly? Is there enough of it to train meaningful models? Can we access it in the format needed? Do we understand its biases and limitations? Can we assess the Data?

A healthcare system fell into this trap spectacularly. They decided to implement AI for patient readmission prediction. They had decades of patient data, millions of records. Surely that was sufficient.

Except the data existed across seven different systems that did not integrate cleanly. Patient identifiers were not consistent. Diagnosis codes had changed over time. Documentation completeness varies by department and physician. Social determinants of health, which significantly predicted readmission, were not systematically captured.

They spent eighteen months just preparing the data before they could meaningfully train models. The data preparation cost more than the AI development itself. Had they honestly assessed data readiness before committing to the project, they might have chosen a different approach or allocated budget very differently.

The data foundation problem has multiple dimensions that organizations often do not appreciate until they are deep in implementation.

Quality: Data needs to be accurate and complete. Missing values, inconsistent formats, and errors all degrade AI performance. Most operational data is collected for operational purposes, not analytical perfection. The standard that worked fine for running the business often is not adequate for training AI.

Volume: Machine learning requires substantial data. How much depends on the problem, but it's usually more than organizations estimate. A few hundred examples aren't enough. A few thousand might not be enough. Tens of thousands start to be meaningful for simpler problems, hundreds of thousands, or millions for complex ones.

Labeling: Supervised learning requires labeled data, examples where you know the right answer. Creating labeled datasets is time-consuming and expensive. Organizations often discover too late that they have lots of data but truly little of it is labeled in ways useful for their AI initiative.

Bias: Data reflects the processes and decisions that created it. If those processes were biased, the data is biased. AI trained on biased data perpetuates and often amplifies those biases. Detecting and correcting bias in historical data is extraordinarily difficult.

Accessibility: Data might exist but be trapped in legacy systems, spread across incompatible databases, or restricted by privacy and regulatory constraints. Having the data and being able to use it for AI are different things.

A retail bank discovered all these issues simultaneously when they attempted to build AI for credit risk assessment. They had decades of loan data, but it was inconsistent across product lines, missing key variables, biased by historical lending practices they were now trying to move away from, and split across systems that could not easily share information.

What they thought would be a six-month AI development project became a two-year data infrastructure overhaul. They succeeded eventually, but only after acknowledging that their data foundation was not remotely ready for what they were attempting.

The honest question organizations need to ask is not "Do we have data?" It is "Do we have data good enough for this specific AI application?" And the answer needs to come from someone who understands both the data intimately and AI requirements realistically, not from someone optimistically assuming it will all work out.

Many organizations should pause their AI initiatives and invest in data infrastructure first. Build the foundation, then build the AI. This feels slow and unglamorous. It is hard to get excited about data quality initiatives. Skipping over this step and diving straight into AI is building on sinking sand.

SIN 4: UNDERESTIMATING CULTURAL RESISTANCE

Technology changes are straightforward. Human changes are not.

Organizations approach AI as a technology challenge: building the model, integrating the systems, training the users. They allocate budget for software, hardware, and consulting. They create technical project plans with milestones for development, testing, and deployment.

What they do not adequately plan for is that AI fundamentally changes how work gets done, how decisions get made, and often, whether certain roles continue to exist. These changes threaten people's autonomy, expertise, and job security. Threatened people resist ways that undermine even technically successful implementations.

A credit union implemented AI for loan approval decisions. Technically, it worked well. The models were accurate, the integration was clean, the system was reliable. However, loan officers, who had built careers in relationship banking and personal judgment, resented being overruled by algorithms.

They could not openly refuse to use the system. That would be insubordination. Instead, they found subtle ways to work around it. They would spend extra time explaining to customers why the AI decision was wrong and how they would have decided differently. They would route borderline applications to senior officers who could override the AI. They would emphasize to customers that they were "more than just algorithms."

The system's recommendations were being undermined by the very people supposed to implement them. Usage metrics looked good because the system was being queried, but business impact was minimal because the recommendations were often ignored or contradicted.

Management treated this as a training problem. They ran workshops on "working with AI." They created incentives tied to system usage. They communicated the strategic importance. None of it worked because they were addressing symptoms, not cause.

The cause was that loan officers felt devalued. Their professional identity was built on judgment and expertise. The AI implied their judgment was replaceable, their expertise irrelevant. No amount of training or communication could overcome that fundamental threat.

The credit union eventually succeeded, but only after redesigning the process so loan officers were positioned as experts who supervised AI, not as workers replaced by it. They framed the AI as a tool that freed officers from routine decisions so they could focus on complex cases and relationship building. This narrative preserved professional identity while still leveraging the technology.

Cultural resistance manifests in countless ways:

Managers who quietly continue using old processes because they trust them more than the AI.

Employees who comply technically but without enthusiasm, doing the minimum required to avoid trouble.

Departments that find reasons why the AI does not work in their unique situation, carving out exceptions that gradually swallow the implementation.

Subtle information gatekeeping where people don't actively sabotage but also do not go out of their way to help the AI succeed.

Stories that circulate AI failures, emphasizing every mistake while ignoring every success.

None of these are malicious. It is human. People protect their interests, preserve their status, and resist changes that threaten their position. Organizations that ignore these dynamics in favor of focusing purely on technical implementation discover that perfect technology cannot overcome human resistance.

The mistake is not attempting transformation. The mistake is expecting technology deployment to accomplish transformation without addressing the human and cultural dimensions that determine whether people actually use the technology as intended.

A hospital system implementing AI diagnostic support learned this the hard way. Physicians resented the implication that they needed algorithmic assistance. Many quietly ignored the AI recommendations. When forced to engage with the system, they documented reasons for overriding it, creating a paper trail of "AI failure" that undermined the initiative politically.

The technology was sound. The physicians were skilled. The implementation treated the physicians as passive recipients of modern technology rather than as professionals whose autonomy and judgment were being challenged. That framing guaranteed resistance.

Successful AI implementations start with honest acknowledgment that this is not just a technological change. It is an organizational Transformation that affects power, status, and identity. It requires change management that is at least as sophisticated as the technical implementation. Most organizations vastly underinvest in the human side relative to the technical side, then wonder why technically successful systems fail operationally.

SIN 5: LACK OF CLEAR SUCCESS METRICS

Ask most organizations implementing AI how they will know if they have succeeded, and you will get vague answers.

"Improved efficiency." "Better customer experience." "Enhanced decision-making." These are not success metrics. They are aspirations. They cannot be measured objectively or evaluated honestly.

Without clear success metrics defined before implementation begins, organizations drift into self-congratulation or endless disappointment. They can claim success based on cherry-picked indicators or they can declare failure because expectations were impossibly vague.

A telecommunications company invested $12 million in AI-powered network optimization. When the CTO was asked by the board how they were measuring success, he listed five different metrics: network uptime, customer satisfaction, operating costs, incident resolution time, and capacity utilization.

Sounds comprehensive. Except none of these metrics had baseline measurements, target improvements, or periods. "Better than we were" is not a success metric. It is a rationalization framework.

Two years later, the CTO could point to improvements in some metrics while acknowledging challenges in others.

Overall, had it succeeded? The question sparked a twenty-minute discussion that concluded "it depends how you look at it." That ambiguity had been baked in from the start by failing to define success clearly.

Clear success metrics require specificity:

What exactly are we measuring? Customer satisfaction is vague. Net Promoter Score at 60 or above is specific. Operating costs are vague. Reducing processing costs by 15% per transaction is specific.

What is the baseline? You cannot evaluate improvement without knowing where you started. Baseline measurement forces honest assessment of current performance.

What is the target? How much improvement constitutes success? This needs to be realistic but meaningful. A 2% improvement might not justify millions in investment. A 50% improvement might be unrealistic.

What is the timeframe? When should we expect to hit targets? This creates accountability and prevents indefinite "it's still early" rationalization.

A logistics company did this properly. They were implementing AI route optimization. Their success metrics were specific: reduce fuel costs by 8% within twelve months, reduce overtime costs by 12% within twelve months, maintain or improve delivery reliability as measured by on-time percentage.

These metrics were ambitious but achievable based on pilot data. They had clear baselines from current operations. They had specific timeframes and critically they agreed on these metrics before implementation began, not after.

When they evaluated the initiative for twelve months, the assessment was straightforward. Fuel costs were down 7.5%, just shy of target. Overtime costs were down 15%, exceeding target. Delivery reliability had improved slightly. Clear success, with honest acknowledgment of where they came up short and where they exceeded expectations.

The discipline of defining success metrics forces important conversations. If you cannot agree on what success looks like, you should not proceed. If the metrics you can agree on do not justify the investment, you should not proceed. If you are uncomfortable committing to specific targets, you aren't ready.

Organizations skip this discipline because it is uncomfortable. Vague aspirations are safe. Specific, measurable commitments create accountability. But that accountability is precisely what is needed to prevent AI initiatives from becoming exercises in expensive ambiguity.

SIN 6: THE "PILOT PURGATORY" TRAP

Pilots are supposed to assess feasibility, identify issues, and inform scale decisions. Instead, many organizations use pilots as perpetual proof that they are "doing AI" without ever committing to full implementation.

The pattern is familiar. Launch a pilot. Get mixed results. Learn some things. Decide to do another pilot incorporating those learnings. Get different mixed results. Launch a third pilot to address newly identified issues. Five years later, you are still piloting, still learning, still preparing for the eventual scale that never comes.

This is pilot purgatory, and it is surprisingly common.

An insurance company has been piloting AI claims processing for four years. Each pilot reveals issues: data quality, process integration, user acceptance. Each new pilot attempts to address those issues. Progress happens, but so slowly that by the time one set of challenges is resolved, the technology landscape has shifted, and new possibilities emerge that require fresh evaluation.

They are learning, certainly. They are gaining expertise. They are not transforming their business. The pilots have become an end in themselves rather than a means toward production deployment.

Pilot purgatory happens for several reasons.

Risk aversion: Organizations are more comfortable with limited pilots than full commitment. Pilots feel safe. They are experiments. Failure is expected. Full deployment is commitment. Failure is embarrassing.

Perfectionism: There is always something that could be improved. The AI could be more accurate. The integration could be smoother. The user experience could be better. If you wait until everything is perfect, you wait forever.

Organizational inertia: Pilots do not require complex decisions about resources, roles, or processes. Scaling does. It is easier to run another pilot than to make hard choices about implementation.

Lack of executive commitment: True executive sponsorship means pushing through obstacles toward scale. Passive executive support means tolerating perpetual piloting without demanding progress.

The antidote is establishing clear criteria for pilot success upfront. Define what results would justify scaling. Set a timeframe for the pilot-to-production decision. Commit that if pilot results meet criteria, you will scale, and if they do not, you will stop rather than perpetually pilot.

A manufacturing company did this well. Their AI quality control pilot had explicit success criteria: detect defects at 95% accuracy or better, false positive rate below 3%, integration with existing workflow without slowdown. Timeframe: six months.

At six months, they evaluated against those criteria. Accuracy was 94%, just below target. False positives were at 2.5%, better than target. Integration had minor workflow impacts that needed addressing.

Decision: good enough to proceed with scaling, with parallel focus on pushing accuracy above 95%. They gave themselves three months to hit that final metric while beginning broader deployment.

This worked because they'd committed to a decision timeline and clear criteria. The pilot was not perpetual exploration. It was time-bound evaluation leading to a committed decision.

Pilots should have expiration dates. If the criteria are not met by that date, either stop the initiative or be honest that you're not ready and need to address fundamental gaps before piloting again. What you should not do is drift into another pilot without honest evaluation of why the previous one did not lead to scale.

Organizations in pilot purgatory often need external pressure to break the pattern. A new executive who demands production results. A competitor who scales AI successfully. A board that questions why years of piloting have not produced business impact.

That pressure is painful. Better to avoid pilot purgatory from the start by treating pilots as what they should be: time-bound tests leading to clear scale-or-stop decisions, not perpetual exploration programs.

SIN 7: ABDICATION OF LEADERSHIP RESPONSIBILITY

The deadliest sin, the one that often enables all the others, is when leadership delegates AI to specialists and abdicates responsibility for strategic direction and accountability.

This takes a predictable form. The board or executive team decides AI is important. They authorize budget and mandate progress. Then they hand it to the CIO or Chief Digital Officer or a specially created AI task force and mostly step back.

Updates come through presentations that emphasize progress and downplay problems. Leadership asks questions but does not push hard when answers are technical and complex. The initiative proceeds with leadership awareness but not leadership ownership.

This abdication creates vacuum. No one with real authority is tracking whether the initiative solves business problems versus just deploying technology. No one is empowered to stop the project if it is not working. No one is holding people accountable for results versus activity.

A retail chain's CEO approved $15 million for implementation of AI and tasked the Chief Information Officer with execution. The CEO attended monthly steering committee meetings but deferred to technical experts on decisions. When problems emerged, the CIO assured him they were normal implementation challenges.

Eighteen months in, the initiative had delivered minimal value. The CEO felt frustrated but also unclear about what had gone wrong or whose responsibility it had been to prevent it. He had funded it. He'd monitored it but not owned it.

Leadership responsibility in AI transformation means several specific things:

Defining business outcomes: Leaders must articulate what business problems AI is solving and what success looks like. This cannot be delegated to technical teams who optimize technical metrics.

Maintaining strategic oversight: Regular, substantive engagement with how the initiative is progressing against business objectives, not just technical milestones.

Making challenging decisions: When kill criteria are hit, when pivots are needed, when additional resources are required, leaders must decide. These decisions can't be studied endlessly or delegated.

Accountability: Someone senior must be personally accountable for results. Not for trying, not for effort, but for actual business outcomes.

Resource allocation: AI transformation requires investment not just in technology but in data infrastructure, talent development, change management, and organizational redesign. Leaders must make these resource decisions, not just approve a lump-sum technology budget.

When leadership abdicates, these responsibilities fall to mid-level managers who lack authority to secure resources, make strategic trade-offs, or hold peers accountable. The initiative becomes one more corporate program that everyone knows is supposed to be important, but no one can drive to a successful completion.

A pharmaceutical company's AI initiative succeeded largely because their COO owned it personally. She did not understand the technical details, but she understood the business problems they were solving and held weekly reviews focused on business metrics, not technical progress.

When integration challenges emerged that required other departments to reprioritize resources, she made it happen. When budget needed reallocation from other initiatives, she made those trade-offs. When pilot results were mediocre and the team wanted to continue anyway, she insisted on honest evaluation and course correction.

This was not micromanagement. It was engaged leadership refusing to abdicate responsibility for a strategic initiative just because it involved complex technology.

The question is not whether leaders need to understand AI deeply. They do not. The question is whether they're willing to maintain strategic ownership of initiatives that use AI, applying the same rigor and accountability they would apply to any other major transformation.

When they do, success rates improve dramatically. When they do not, when they assume that AI is technical and therefore someone else's problem, failure rates stay stubbornly high.

THE CUMULATIVE WEIGHT OF SIN

Organizations rarely commit just one of these sins. More commonly, they commit several simultaneously, and the effects compound.

They start with technology instead of business problems (Sin 1), believe vendor hype without validation (Sin 2), underestimate data requirements (Sin 3), and ignore cultural resistance (Sin 4). Four sins create a guaranteed failure, yet this exact combination appears constantly.

A financial services firm committed all seven simultaneously. They started with technology exploration rather than business problems. They believed impressive vendor demonstrations without adequate validation. They assumed their data was adequate when it was not. They ignored cultural resistance from relationship managers. They had vague success metrics. They cycled through multiple pilots without scaling and leadership abdicated ownership to IT.

The result was predictable: $22 million spent, minimal value delivered, organizational cynicism about future innovation.

What is tragic is that each of these sins is avoidable. They are not subtle traps requiring sophisticated expertise to recognize. They are obvious mistakes that organizations make anyway because of momentum, optimism, or competitive pressure.

Understanding the seven sins does not guarantee you will avoid them. Organizational dynamics can push these mistakes even when individuals recognize them. But understanding at least creates the possibility of different choices, of pausing when momentum says proceed, of questioning when consensus says accept.

The organizations succeeding with AI are not necessarily smarter or better resourced than those failing. They avoid these seven sins, either through conscious discipline or through lucky circumstances that protected them from their own potential mistakes.

The next chapter explores why these sins are so seductive, examining the cognitive biases and organizational dynamics that lead smart companies to make predictably bad decisions about AI. Understanding what not to do is only half the solution. The other half understands why we're tempted to do it anyway.

CHAPTER 3

WHY SMART COMPANIES MAKE DUMB DECISIONS

The CFO had done everything right on paper.

She had commissioned three separate AI readiness assessments, hired McKinsey to validate the business case, and assembled a steering committee of department heads who'd all nodded enthusiastically through six weeks of presentations. The project had executive buy-ins, a reasonable budget, and a vendor with glowing references. By any rational measure, this should have worked.

Eighteen months later, she sat in a conference room discussing the challenges that they had faced. "I don't understand," she said. "We're not stupid people. We run a multi-billion-dollar operation. We have successfully integrated three acquisitions, modernized our entire ERP system, and navigated two economic crises. "But this we couldn't figure out."

How did they get this so spectacularly wrong?

The answer is not comforting. Intelligence does not protect you from bad AI decisions. Experience does not exist either. In fact, success in traditional business transformation can make you more vulnerable, not less, because it creates a dangerous overconfidence in your ability to spot what is different this time.

What makes smart companies make bad decisions about AI is not a lack of information or capability. It is something more fundamental: a predictable set of cognitive traps that distort judgment at exactly the wrong moments and once you understand these patterns, you start seeing them everywhere.

THE BOARDROOM CONTAGION: FOMO AT THE EXECUTIVE LEVEL

Start with the pressure cooker, which is modern executive decision-making.

You are at the quarterly board meeting. The technology committee chair, who just returned from a conference in Singapore, mentions that your largest competitor announced an AI initiative. Two smaller competitors have already implemented something. An analyst report you saw last week suggested that companies not adopting AI risk becoming obsolete within five years. Your Chief Digital Officer has been lobbying for budget. A vendor has been persistently helpful, sending research papers and offering to present to the board at no cost.

This is how it begins. Not with careful strategy, but with accumulating anxiety.

Fear of Missing Out was not supposed to apply to Fortune 500 companies. It is a teenager's affliction, a social media phenomenon. Except it turns out that executives are just as susceptible to status anxiety and competitive panic as anyone else, more so because the stakes feel existential. When everyone else seems to be moving on AI, staying put starts to feel like strategic negligence.

A manufacturing executive described the dynamic perfectly: "No one wanted to be the person who said 'let's wait and see' when the board was asking what we were doing about AI. That is a career-limiting position. So, we all found reasons why now was the right time, even though six months earlier we had agreed our data infrastructure wasn't ready."

The decision was not made through analysis. It was made through social pressure and the subtle but overwhelming need not be left behind.

What makes this particularly insidious is that FOMO masquerades as prudence. It dresses itself up in the language of competitive positioning and strategic foresight. The board presentation does not say "we're panicking because everyone else is doing this." It says "first-mover advantage" and "digital transformation imperative" and "capturing market share in the AI era." The panic is laundered through PowerPoint into something that looks like strategy.

Research by Deloitte in 2024 found that 64% of executives cited competitive pressure as the primary driver for AI investment, ahead of specific business problems or clear ROI projections. They were making multi-million-dollar decisions to avoid feeling left out.

And here is what compounds the problem: once you have announced an AI initiative, once you've committed budget, hired consultants, and told investors you're embracing innovation, the pressure to continue intensifies. Admitting six months in which you should not have started becomes almost impossible. The sunk cost fallacy kicks in, and what began as FOMO transforms into something even more dangerous: the need to justify what you have already done.

THE MAGIC BULLET FALLACY: WHEN AI BECOMES THE ANSWER TO EVERYTHING

There is a particular kind of organizational magical thinking that AI seems to trigger.

I have watched it happen in real time. A company has struggled for years with poor sales forecast. They have tried new processes, better training, different compensation structures. Nothing quite works. Then someone mentions AI, and suddenly the room changes energy. Yes, that is it. Machine learning can analyze historical patterns. It can factor in market conditions. It can eliminate human bias. AI will solve the forecasting problem.

Except the forecasting problem was not actually a data analysis problem. It was a problem of misaligned incentives. Regional managers inflated their projections to secure resources, then sandbagged actual forecasts to beat expectations. No amount of algorithmic sophistication was going to fix a system where everyone benefited from inaccurate numbers.

The magic bullet fallacy operates on a simple but seductive premise: technology can solve problems that are fundamentally about people, culture, and organizational dysfunction. It is appealing because technology is clean. You can buy it, implement it, measure it. It does not require confronting uncomfortable truths about how your organization works.

A financial services company spent eighteen months building an AI system to improve customer service response times. The system was technically impressive. It could route inquiries, predict customer needs, and suggest responses. Response times barely improved because the real problem was that customer service was treated as a cost center with chronic understaffing and high turnover. The AI just gave burned-out representatives another system to learn while they juggled too many customers.

What I have noticed is that the magic bullet fallacy is most powerful in organizations that are otherwise well-managed. Competent companies develop confidence in their ability to execute solutions. When they identify a problem, they mobilize resources, bring in expertise, and fix it. This creates a mental pattern: problem → solution → implementation → success.

AI exploits this pattern because it presents itself as a comprehensive solution. The vendors certainly pitch it that way. Transform your business." "Unlock unprecedented insights." "Stay ahead of the competition." The language is all-encompassing, suggesting that AI is the thing you have been missing, the key that unlocks everything else.

But organizational dysfunction is not a technical problem. If your processes are broken, AI will automate broken processes. If your data is politicized, AI will systematize bad data. If your teams don't trust each other, AI will just give them new things to argue about. Technology amplifies what is already there, good, or bad.

A hospital system learned this the expensive way. They implemented AI for patient flow management, convinced it would solve their chronic emergency room overcrowding. The AI was smart. It could predict admission rates, optimize bed allocation, and flag bottlenecks. But the overcrowding persisted because the significant issues were territorial disputes between departments, inadequate discharge planning, and a culture where hoarding beds were rewarded. The AI just gave everyone data to support their existing positions.

The brutal truth: if you can't solve a problem with better management, clearer accountability, and honest communication, you certainly cannot solve it with AI. Technology does not fix culture. It reveals it.

CONFIRMATION BIAS: SEEING ONLY WHAT SUPPORTS THE DECISION

Once significant budget has been committed, something subtle and dangerous happens to organizational perception.

You start cherry-picking evidence.

The pilot shows mixed results, some promising indicators but significant operational challenges. In a rational evaluation, this might prompt reconsideration. When millions are committed, when the initiative has been announced, when careers are now attached to their success, the mind does something clever: it starts weighing the promising indicators more heavily and finding explanations for the challenges.

"These implementation issues are normal at this stage."

"The real benefits will emerge once we scale."

"The team is still learning; we need to give it more time."

None of these statements are necessarily false. That is what makes confirmation bias so effective. You are not lying; you are just giving more credence to the evidence that supports continuation and less weight to the warning signs.

A telecommunications company exhibited this pattern almost textbook-perfectly. Their AI initiative for network optimization showed early problems: the model's predictions were less accurate than their existing system, and engineers were spending more time correcting AI recommendations than the system saved. But the project had been officially announced as part of their digital transformation strategy. The CEO had mentioned it in an earnings call.

So, the narrative shifted. The accuracy problems became "the learning curve." The additional engineering time became "an investment in building organizational capability." The fact that the old system worked better became "legacy thinking that resists innovation."

By month nine, the steering committee meetings had an almost surreal quality. Everyone knew the initiative was not working, but the language used in discussion carefully avoided saying so directly. Reports emphasized improvements in model architecture while downplaying continued accuracy issues. Success metrics were quietly adjusted. The definition of "fully operational" became more flexible.

This is confirmation bias at the institutional level, and it is particularly dangerous because it is collective. It is not one person deluding themselves: it is a room full of smart people unconsciously coordinating to avoid confronting uncomfortable reality.

Research on organizational decision-making suggests this is inevitable once commitment exceeds a certain threshold. Psychologists call it the escalation of commitment: the tendency to invest more resources in a failing course of action because you have already invested so much. In business, we call it "throwing good money after bad," but we rarely recognize it while we are doing it.

The warning signs get reinterpreted. The dissenting voices get dismissed as resistant to change. The straightforward evidence that something isn't working gets filed away as "feedback for the next iteration." The project continues because stopping it would require admitting the initial decision was wrong.

A manufacturing company continued their AI quality control project for fourteen months after it became clear the system could not reliably identify defects. Why? Because they had already invested $3.2 million and told their board it would transform production quality. The cognitive dissonance of admitting failure was too painful, so they kept iterating, hoping the next version would vindicate the original decision. It never did. They finally cancelled it when a major client threatened to switch suppliers after receiving a shipment with defects the AI had missed.

The most dangerous phrase in AI transformation is: "We have come too far to turn back now."

THE EXPERTISE GAP: WHEN NO ONE KNOWS ENOUGH TO ASK THE RIGHT QUESTIONS

Here is a scenario that plays out in some variation everywhere:

The vendor's technical team presents the proposed solution. They are credible, knowledgeable, articulate. They walk through the architecture, the machine learning approach and the expected performance metrics. They answer questions confidently. At the end, the CFO looks around the room. "Any concerns?" Heads shake. The CIO nods thoughtfully. "Sounds solid."

No one wants to admit they did not fully understand the technical explanation.

This is the expertise gap in action, and it is particularly acute with AI because the technology exists at the intersection of multiple specialized domains. You need to understand the business context, the data infrastructure, the algorithmic approaches, the implementation complexity, and the change management challenges. Very few people have all that knowledge. Most organizations have no one who does.

But admitting ignorance in a senior leadership setting carries social cost. It signals you are not keeping up, not equipped for modern business, not the right person for your role. So, people stay quiet when they are confused. They ask surface-level questions that sound engaged without revealing gaps in understanding. They delegate the technical evaluation to IT without recognizing that IT does not understand the business context.

The result is decisions made through deference rather than comprehension.

A retail executive discussed approving a $2.7 million AI customer segmentation project: "I'm not technical. When the vendors showed their algorithm performance metrics, I had no frame of reference for whether 87% accuracy was good or terrible for this application. I looked at our CTO, he seemed satisfied, so I approved it. Turns out 87% wasn't nearly good enough for what we needed."

The expertise gap is not about technical knowledge. It is also about understanding what questions matter.

When evaluating a traditional IT project, executives have decades of pattern recognition to draw from. They know that scope creep is dangerous, that integration is complex, that user adoption is critical. They have seen enough implementations to smell trouble.

With AI, those patterns do not fully apply. The questions that matter are different: Is our data suitable for this approach? How will the model behave at the edges of its training? What happens when market conditions shift? How do we validate that it is working? Most executives do not yet have the experience base to generate these questions naturally.

So, they rely on vendors and consultants, who have their own incentives. The vendors are not being dishonest, exactly, but they're presenting the best case, not the full spectrum of possibilities. They emphasize what technology can do, not what it cannot. They show successful case studies, not the messy reality of most implementations.

An insurance company approved an AI claims processing system based on vendor demonstrations showing remarkable accuracy. What they did not know to ask: those demonstrations used clean, well-structured claims data from ideal scenarios. Real-world claims are messy, incomplete, and often contradictory. The system that performed at 94% in demos struggled to break 70% in production.

The expertise gap creates a fundamental power asymmetry. The people with the knowledge to evaluate the technology are not usually the ones making strategic decisions. The people making strategic decisions do not have the knowledge to properly evaluate technology. And the bridge between them, the people selling and implementing the technology, have incentives that do not perfectly align with objective assessment.

This is not anyone's fault. It is a natural characteristic of emerging technology. But it creates an environment where bad decisions are structurally likely, where organizations commit to approaches, they do not fully understand because the alternative is admitting they are out of their depth.

GROUPTHINK: HOW CONSENSUS KILLS CRITICAL THINKING

Watch how AI decisions typically get made in organizations. A working group is formed, usually people identified as forward-thinking, innovative-oriented, positive about modern technology. This makes sense, you want enthusiastic people on an innovative initiative. They meet regularly, develop shared understanding of the opportunity, build momentum and camaraderie. By the time they present to senior leadership, they are a unified front, speaking with one voice about the strategic imperative and implementation plan.

This is groupthink and it is how organizations silence the doubts that might save them.

Groupthink does not require a domineering leader or deliberate suppression of dissent. It emerges naturally when groups value consensus and cohesion. People start self-censoring doubts to maintain group harmony. Contrarian positions get softened or withdrawn. The range of options considered narrow. And gradually, without anyone noticing, the group converges on a perspective that feels like objective truth but is just collective agreement.

AI initiatives are particularly vulnerable to groupthink because they involve uncertainty and complexity. When you do not have perfect information, social proof becomes a powerful guiding principle: if everyone else thinks this is right, it is. The group's emerging consensus substitutes for the objective analysis that is difficult to perform.

This unfolded at a pharmaceutical company. The AI working group spent three months evaluating drug discovery applications. By month two, they had become advocates, excited about the possibilities, bonding over their shared understanding of this important initiative. When a senior scientist raised concerns about the quality of their compound data, the group response was polite but dismissive: "We have addressed that in our data preparation workstream." The scientist did not push back. The concern disappeared.

Six months into implementation, the data quality issues the scientist had identified became critical problems. By then the group had disbanded, the decision was made and raising it would require admitting everyone had missed something important.

Groupthink is especially dangerous in cross-functional teams, which are exactly the structure most organizations use for AI initiatives. You bring together IT, operations, finance, and business units to ensure diverse perspectives but if the group gels too quickly, that diversity gets subsumed into collective thinking. Everyone knows the other functions have raised no major objections, so their own hesitations start to feel like misunderstandings or resistance to change.

A logistics company formed a diverse steering committee for their AI route optimization project. IT, operations, finance, customer service, all represented. In the early meetings, people raised different concerns: IT worried about system integration, operations about driver acceptance, finance about ROI timelines. As the project gained momentum, these concerns were "worked through" in the committee. By month four, everyone was on board.

Except they were not, not really. Operations still worried about driver acceptance but felt like they'd lost that argument. IT still had concerns of integration but did not want to be seen as blocking progress. The public consensus was an illusion. The private doubts remained, unspoken but real.

When the system was launched, drivers rebelled. The routes the AI generated violated informal knowledge about traffic patterns, customer preferences, and practical constraints that the algorithm could not capture. The drivers had worked these routes for years; the AI had worked them for zero days. The result was predictable chaos.

Post-mortem interviews revealed that everyone had seen various parts of this coming. But in the consensus-seeking environment of the steering committee, no one had wanted to be the person who stopped progress.

The antidote to groupthink is designated dissent, someone whose explicit role is to challenge assumptions and poke holes in logic. Most organizations do not create this role for AI initiatives. It feels negative, counter to the innovation-forward culture they are trying to build. So, the groupthink proceeds unchecked, and decisions that should have been challenged sail through on a wave of collective agreement.

THE VENDOR-CLIENT DYNAMIC: INFORMATION ASYMMETRY AND MISALIGNED INCENTIVES

Let us talk about the uncomfortable truth in the room: the people who know the most about whether your AI initiative will work are also the people making money from it proceeding.

This is not an indictment of vendors. Most are professional, many are genuinely trying to deliver value, and the technology they sell often does what they say it does. But the incentives are structurally problematic, and pretending otherwise is naive.

A vendor's primary objective is to close the deal and deliver the project. Your primary objective should be determining whether the project is right for your organization. These objectives align sometimes, but not always. In the gap between sometimes and always, expensive mistakes happen.

The information asymmetry is stark. Vendors have implemented their technology dozens or hundreds of times. They know intimately what works, what does not, what makes clients successful, what leads to struggles. You are doing this once, maybe twice in your career. The knowledge imbalance is massive.

This would not matter if incentives were perfectly aligned, but they are not. A vendor can deliver exactly what was promised and still have your project fail if the promise was wrong for your context. They can be honest about capabilities while being optimistic about your readiness. They can genuinely believe in their technology while underestimating your implementation challenges.

And sometimes, though no one likes saying it directly, vendors knowingly oversell.

A financial services company spent $4.8 million on an AI fraud detection system that the vendor knew would require data integration work beyond what the company had budgeted or planned for. The vendor did not lie about this, exactly. They mentioned it in the technical documentation, but they didn't emphasize it in the sales process, and when implementation revealed the extent of the integration challenges, they expressed surprise that the client hadn't understood the requirements.

Had they been deceptive? Not legally. They had documented everything. They had managed the sales conversation to minimize concerns and maximize perceived value, which is what good salespeople do. The client signed a contract for a solution that, as sold, could not work without significant additional investment.

The vendor dynamic gets more complex with consultants and systems integrators, who often play a dual role as advisors and implementers. They are helping you decide what to do while also hoping to be hired to do it. This creates subtle but real conflicts.

I have seen proposals where the consulting firm's "objective" assessment of AI opportunities somehow always concluded that a substantial implementation project was needed, ideally one matching the firm's capabilities. Coincidence? Maybe. Or maybe when you are a hammer, everything looks like a nail, and when you are an AI consultant, every business problem looks like an AI opportunity.

The power dynamic matters too. Vendors and consultants often have more experience in negotiation and contracting than their clients have in buying these services. They know which terms to push for, which risks transferring, which ambiguities benefit them. By the time you have learned these lessons, the contract is signed and the project is underway.

A manufacturing company discovered eighteen months into their AI implementation that "production-ready system" in their contract had a definition different from what they had assumed. To the vendor, it meant the code was deployable. To the manufacturer, it meant the system could run their production line. The gap between these definitions represented about $1.2 million in additional work. The contract language technically supported the vendor's interpretation.

None of this makes vendors villains. It makes them rational economic actors in a market where the buyer often does not know enough to be a sophisticated purchaser. But it does mean that relying primarily on vendor guidance to make AI decisions is structurally risky.

The best protection is independent expertise, someone who understands technology but does not benefit from the project proceedings. Except those people are rare and expensive. So, most organizations make do with the expertise they can access, which usually means the people selling them something.

THE TEN COMMON DECISION FAILURES: A PATTERN RECOGNITION GUIDE

After studying hundreds of AI initiatives across industries and continents, certain decision failures emerge repeatedly. Think of these as the greatest hits of AI mistakes, the patterns that smart organizations somehow keep repeating.

Failure 1: Committing Before Understanding

The decision to pursue AI is made in concept before the specific application is identified. The logic goes: AI is important, we need to be doing something with AI, therefore we should fund AI initiatives. This backwards reasoning leads to solutions searching for problems.

A regional bank decided to "invest in AI" as a strategic priority. They formed a committee, hired consultants, and allocated $5 million before identifying a specific business problem to solve. Nine months later they had funded three pilot projects, none aligned with core business priorities, all languishing because no one had urgency about their success.

The decision failure: prioritizing activity over outcome.

Failure 2: Outsourcing Critical Thinking

Hiring consultants to develop your AI strategy can be valuable. Hiring consultants to think through whether you need an AI strategy is abdication. This happens constantly, with organizations outsourcing the "should we?" question to people who make money from the answer being yes.

A healthcare system paid consultants $400,000 to assess AI opportunities. The consultant report identified fifteen potential applications, recommended starting with five, and outlined a three-year roadmap. The organization followed it. Four years later, only one initiative had delivered meaningful value. The problem: the consultant analysis had been thorough on technical feasibility but superficial on organizational readiness and strategic fit.

The decision failure: letting external parties define your priorities.

Failure 3: Ignoring the Prerequisite Work

AI requires data infrastructure, technical capability, and process maturity. Many organizations attempt AI before they have these foundations, assuming they can build them as they go. This is like deciding to run a marathon before learning to jog.

A retail chain pursued AI inventory optimization while still using inconsistent product codes across locations and manual data entry for stock counts. The AI could not overcome the underlying data chaos. Eighteen months and $2.3 million later, they concluded their data was not ready. This could have been determined in week one with honest assessment.

The decision failure: skipping past the unsexy foundation work.

Failure 4: Confusing Pilot Success with Scalability

A controlled pilot in an optimal environment often succeeds. This gets interpreted as validation that the approach works generally. However, pilots are carefully managed experiments, often with hand-picked data, dedicated resources, and forgiveness for imperfection. Production is none of those things.

A telecommunications company piloted AI customer service in one call center with their most tech-savvy agents, cleanest customer data, and most straightforward queries. It worked beautifully. Scaling to all call centers revealed that the pilot environment had been atypical. Average agents struggled with the system, data quality varied wildly, and complex queries defeated the AI. What worked at pilot scale failed at production scale.

The decision failure: treating proof of concept as proof of readiness.

Failure 5: Underestimating Integration Complexity

The AI model is often the easy part. Connecting it to existing systems, ensuring data flows correctly, handling edge cases, training users, all of this is where implementations flounder. Organizations routinely budget and plan as if integration will be straightforward.

An insurance company built a brilliant AI claims assessment model but integrated it with their policy management system, claims database, and agent workflow tools required eighteen months of additional work and cost triple the original model development budget. The vendor had delivered exactly what was promised. The client had just underestimated everything around it.

The decision failure: focusing on the AI component while ignoring the system.

Failure 6: Succumbing to Sunk Cost Pressure

Once significant money and reputation are committed, stopping becomes psychologically difficult even when continuation is irrational. Organizations pour additional resources into failing initiatives because the alternative is admitting failure.

A logistics company spent $6 million on AI route optimization over two years. By month eighteen it was clear the system could not reliably beat their existing approach. But they had told their board it was strategic, announced it publicly, and restructured operations around its planned deployment. So, they spent another $3 million trying to salvage it before finally, quietly, shutting it down. The last $3 million bought nothing except delay in accepting reality.

The decision failure: letting past commitment drive future investment.

Failure 7: Mistaking Complexity for Sophistication

There is a seductive quality to complex AI solutions. They feel more impressive, more transformative, more worthy of investment. Complexity is often the enemy of success, particularly in early AI initiatives. Simple solutions are easier to implement, debug, and gain adoption for.

A manufacturing company chose a complex AI quality control system that analyzed dozens of variables in real-time when a simpler statistical process control approach would have delivered 80% of the value with 20% of the complexity. They chose complexity because it felt more innovative. They paid for that choice in implementation delays, user confusion, and maintenance overhead.

The decision failure: optimizing for impressiveness over effectiveness.

Failure 8: Neglecting the Human System

AI implementations are socio-technical transformations, but organizations often treat them as purely technical projects. The human elements, how people will react, what incentives they face, what concerns they harbor, these get addressed superficially if at all.

A bank implemented AI loan processing threatened to displace loan officers. Management was surprised when officers subtly sabotaged the system by identifying edge cases where it struggled and routed those to clients, creating an impression of unreliability. The technology was sound. The implementation ignored basic human psychology about job security and professional identity.

The decision failure: treating people as passive recipients of change.

Failure 9: Optimizing for the Wrong Metric

What you measure shapes what you get. Many AI initiatives optimize for metrics that do not actually align with business value. Model accuracy becomes the goal when business impact should be. Deployment becomes the milestone when adoption should be.

A customer service AI was judged successful because it managed 60% of inquiries automatically. What transpired is that customer satisfaction dropped because the 40% that got escalated to humans were the complex, frustrated customers, and agents now had less context to help them. The metric said success. The customer experience said otherwise.

The decision failure: measuring activity instead of outcome.

Failure 10: Believing in Your Own Narrative

Organizations develop stories about their AI initiatives that become increasingly detached from reality. The narrative is that it is going well, transforming the business, positioning for the future. The reality is that it is struggling, delivering marginal value, and creating as many problems as it solves.

A professional services firm maintained for two years that their AI document analysis tool was revolutionizing their workflow. The narrative was in presentations, annual reports, client pitches. The reality: most staff still used the old manual process because the AI was unreliable. The real reality was that the narrative had become institutional truth and challenging it was career-limiting.

The decision failure: confusing what you say with what is true.

A FRAMEWORK FOR BETTER DECISIONS: THE FIVE CRITICAL QUESTIONS

Avoiding these traps is not about being smarter. It is about being more honest and more systematic in how you evaluate AI decisions. Here is a framework that cuts through the cognitive biases and forces clearer thinking.

Question 1: What specific business problems are we solving, and how will we know if we've solved it?

If you cannot articulate the problem in plain language and define what success looks like in measurable terms, you are not ready to invest in a solution. "Improving efficiency" is not specific enough. "Reducing invoice processing time from five days to two days while maintaining accuracy above 99.5%" is.

Force specificity. Demand clarity on metrics. Insist on understanding current performance and target performance. If the problem statement is vague, the solution will be too.

Question 2: Why is AI the right approach for this problem versus other alternatives?

AI is not the only tool in the toolkit. Sometimes better process design works. Sometimes more staff work. Sometimes clearer accountability works. AI should be chosen because it is genuinely the best option, not because it is fashionable.

This question forces comparison. What would a non-AI solution cost and deliver? What have others tried? Why did their approaches succeed or fail? If you cannot clearly articulate why AI is superior to alternatives, you probably haven't thought it through adequately.

Question 3: Do we have the foundational capabilities this requires, or can we build them?

Be brutally honest about data quality, technical infrastructure, organizational readiness, and cultural fit. Most organizations overestimate their readiness because no one wants to admit they are not prepared.

Create a checklist: data availability, data quality, technical skills, integration requirements, management capacity, executive commitment. Rate yourself honestly on each. If you are weak in multiple areas, pause and build foundations first.

Question 4: What are the realistic failure modes, and how would we recognize them early?

Every AI initiative can fail in predictable ways. What are yours? Model does not perform accurately enough. Integration proves too complex. Users do not adopt it. Data quality issues emerge. Costs overrun budget. Timeline slips repeatedly.

Define these upfront and then create early warning indicators for each. If we are heading toward cost overrun, what would we see in month three? If users will not adopt it, what signals would appear during pilot? Build in checkpoints where you evaluate whether failure modes are emerging.

Question 5: Who benefits from us saying yes, and what would they say if we paid them to argue no?

This is the red team question. Your vendors benefit from yes. Your consultants benefit from yes. Your innovation team benefits from yes. They are not lying to you, but their perspective is shaped by their incentives.

Flip it around. What would your vendor's competitor say about their approach? What would a consultant whose expertise is in alternatives to AI say about your situation? What would a skeptical board member say about your readiness? Actively seek the contrary view from someone who has no stake in the decision going forward.

PUTTING IT INTO PRACTICE: THE DECISION REVIEW PROCESS

Frameworks are useless if they are not actually used. Here is a practical process for reviewing major AI investment decisions that incorporates these questions and protects against cognitive traps.

Step 1: The Isolation Test (Week One)

Before any group deliberation, have key decision-makers independently write answers to the five critical questions. No collaboration, no discussion. Just individual assessment.

This prevents groupthink from forming early. You want to know where people actually stand before social dynamics influence their positions. Collect these written responses and look for patterns. Where is there consensus? Where is there disagreement? The disagreements are often where the important questions lie.

Step 2: The Skeptic's Review (Week Two)

Designate someone to argue against the initiative. This is not devil's advocate, where someone half-heartedly raises objections while everyone knows they do not mean it. This is genuine critical analysis by someone whose job is to find flaws.

Give them the proposal, the business case, the technical assessment. Ask them to identify the weakest assumptions, the most likely failure modes, the least convincing parts of the argument. They present this to the decision group not as devil's advocate but as legitimate analysis that deserves serious response.

Step 3: The Vendor Reality Check (Week Three)

If vendors are involved, conduct a structured reference check that goes beyond the glowing testimonials they provide. Ask them for contacts at three implementations that struggled. Every vendor has them. What went wrong? What would they do differently? What surprised them?

Also conduct independent research. Find companies in your industry that attempted similar initiatives. What was their experience? What did they learn? LinkedIn is useful for this; people are often surprisingly candid about AI initiatives that did not work out.

Step 4: The Prerequisite Audit (Week Four)

Before approving budget, audit whether you have the foundations needed. Bring in someone with AI implementation experience who doesn't benefit from the project proceeding. Have they assessed data readiness, technical capability, and organizational preparedness?

This should be a clear pass/fail assessment. If you fail, address the gaps before proceeding. The prerequisite work is never as exciting as the AI implementation, but it is usually more important.

Step 5: The Commitment Checkpoint (Week Six)

Define upfront what success looks like at 90 days, six months, and one year. Define the metrics, the milestones, and the kill criteria. What would have to be true for you to stop this initiative?

Write this down. Make it formal. Review it quarterly. This is your protection against sunk cost fallacy and confirmation bias. When you defined success before you were committed, what did you say? Are you meeting those criteria?

THE PSYCHOLOGY OF SAYING NO

The hardest part of better AI decision-making is not the analysis. It is having the organizational courage to say no when the analysis suggests you should.

Saying no to AI feels like saying no to the future. It feels conservative, unimaginative, risk averse. In organizations that pride themselves on innovation, it can feel like career suicide.

But saying no to the wrong AI initiative is not saying no to AI generally. It is saying yes to doing it right when the conditions are better. It is saying yes to building foundations first. It is saying yes to learning from others' mistakes rather than repeating them.

The most successful AI implementations I have studied often started with multiple "no" decisions. No, we are not ready yet. No, that use case is not appropriate. No, we need to fix our data first. These organizations developed the discipline to decline initiatives that looked good on paper but would not work in their context.

A regional healthcare provider turned down three AI implementation proposals over eighteen months while they upgraded their data infrastructure and built technical capability. Their leadership faced questions from the board about why competitors were moving faster. They held firm. When they finally implemented AI for patient scheduling, it worked smoothly, delivered clear ROI, and created organizational confidence in their ability to do this well.

The organizations that struggle are often the ones that could not say no. They said yes to the first plausible pitch, yes to the vendor with the smoothest presentation, yes because everyone else was doing it. They paid for that inability to say no with failed implementations and wasted investment.

Learning to say no effectively is a critical organizational capability. It requires leaders who can distinguish between being cautiously wise and being fearfully resistant. It requires cultures that reward thoughtful skepticism, not just enthusiastic adoption. It requires metrics that value sustainable success over fast activity.

WHAT THIS MEANS FOR YOU

If you're reading this chapter and recognizing your organization in these patterns, that is positive. Recognition is the first step toward different behavior.

Most organizations making these cognitive errors do not realize they're doing it. They are convinced they are being rational and strategic. The fact that you can see the traps means you're better positioned to avoid them.

The practical implication: slow down your AI decision process. Create space for critical thinking. Build in mechanisms that counteract groupthink, challenge assumptions, and force honest assessment of readiness. Make it organizationally acceptable to say "not yet" without that being interpreted as resistance to innovation.

Smart companies make bad decisions about AI not because they are incompetent but because the decision environment is designed to produce these outcomes. Change the environment, change the process, and you change the results.

The CFO whose story opened this chapter eventually figured this out. Her second AI initiative, launched two years after the first one failed, took six months longer to approve because she had implemented a rigorous decision review process. It included independent technical assessment, prerequisite capability building, and clear kill criteria.

That second initiative worked. It delivered measurable ROI, gained user adoption, and created organizational capability for future AI implementation.

The difference was not smarter people or better technology. The difference was better decisions, protected from the cognitive traps that doom most AI transformations before they start.

CHAPTER 4

DECODING THE SUCCESS PATTERN

The CEO of a regional insurance company in Melbourne had a habit of asking uncomfortable questions. Not the aggressive kind that makes people defensive, but the genuine kind that exposes what everyone has been thinking but nobody wants to say.

Six weeks into their AI initiative, he asked one during a steering committee meeting: "If we stopped this project right now, what would we actually lose?"

Silence. The kind that feels thick.

The IT director eventually spoke. "We'd lose the momentum we've built. The team's engagement. The vendor relationship."

"Right," the CEO said. "But what business value would we lose? What customer problem would remain unsolved? What operational inefficiency would continue costing us money?"

More silence.

"That's what I thought," he said, not unkindly. "We've spent six weeks and nobody can articulate what we'd lose by stopping. That tells me we don't actually know what we're building or why it matters."

He did not kill the project. Instead, he did something that felt radical: he paused it. Tell the team to stop all development work for two weeks. Spend that time interviewing claims adjusters, underwriters, and customer service reps. Find out what problems actually need to be solved. Come back with three specific, measurable business outcomes they could commit to delivering.

The team hated it at first. It felt like failure, like admission they had been heading in the wrong direction. When they came back two weeks later, something had shifted. They were not talking about AI capabilities anymore. They were talking about claims processing bottlenecks, customer frustration with policy explanations and underwriter burnout from repetitive risk assessments.

Eighteen months later, their AI initiative had become one of the quieter success stories in the Australian insurance sector. Not because they had built the most sophisticated models or used the fanciest technology. They had figured out early what success actually looked like, and everything they built moved them toward that definition.

This is what separates the successful 20% from everyone else. Not superior technology or larger budgets or smarter teams. They know what they are aiming at. They know what success looks like before they start building towards it and critically, they know what failure looks like soon enough to change course.

THE SUCCESS TRIANGLE: WHEN EVERYTHING CLICKS

Here's something strange about successful AI transformations: they don't feel like technology projects. Talk to people in organizations that have genuinely succeeded with AI, and they'll spend most of their time discussing strategy, culture, and execution. Technology is almost an afterthought, the thing that enabled what they were trying to do rather than the point of the exercise.

I have come to think of successful AI initiatives as requiring three elements working in harmony: Strategy, Execution, and Culture. Miss any one of them, and you are building on sand. Get all three right, and the technology becomes almost straightforward.

Strategy means knowing precisely what business problem you are solving and why AI is the right approach. Not "improve customer experience" but "reduce average claims processing time from 8 days to 3 days by automating the 60% of claims that fit standard patterns." Not "optimize operations" but "decrease inventory carrying costs by 15% through demand forecasting that accounts for seasonal patterns our current system misses."

Successful organizations have strategies you can sketch on a napkin. If you need a 40-slide deck to explain what you're doing and why, you don't understand it yourself.

Execution means building and deploying AI in ways that generate learning even when they do not generate immediate results. The successful 20% fail constantly. They just fail in informative ways. They conduct small experiments, measure what happens, adjust based on what they learn, and scale what works. They kill projects that are not delivering value, not because they are punishing failure but because they are allocating resources to better opportunities.

Organizations that struggle with AI tend to have execution that is either too rigid or too loose. Too rigid, and you are following a plan even when reality screams that the plan is wrong. Too loose, and you are wandering, trying things without learning from them, never building momentum.

Culture is the hardest and most crucial element. It is the difference between AI that people use and AI that generates reports nobody reads. Culture determines whether your claims adjusters will trust the system enough to let it route simple claims automatically, or whether they will insist on reviewing every decision manually, destroying the efficiency gains you were hoping for.

You can't mandate culture change through policy. The

insurance CEO in Melbourne did not send out a memo saying, "Everyone will now embrace AI." Instead, he involved claims adjusters in identifying which parts of their work were tedious and which required human judgment. He let them help design the system. He made it clear that AI was there to manage the boring stuff so they could focus on complex cases that actually needed expertise.

When the three elements of the Success Triangle align, transformation stops wanting to push a boulder uphill. It starts feeling inevitable, natural, like you are finally working with gravity instead of against it.

THE FIRST 90 DAYS: EARLY WARNINGS WORTH HEEDING

A research team at MIT studied 200 AI implementations and found something remarkable: they could predict with 78% accuracy which projects would succeed or fail based entirely on what happened in the first 90 days. Not the technology choices. Not the budget. Not even the team composition, but specific behavioral patterns that emerged early.

The successful projects shared five characteristics in those first three months:

They spent more time defining the problem than exploring solutions.

This sounds obvious until you watch organizations actually do it. Most teams, once given budget and permission to pursue AI, immediately start evaluating platforms and interviewing vendors. The successful ones spend week one talking to the people who actually do the work they are trying to improve. They map current processes in painful detail. They quantify exactly how much time, money, or customer satisfaction is being lost.

A logistics company in Rotterdam spent their entire first month just observing how dispatchers made routing decisions. No coding. No planning. Just watching and asking questions. They discovered that the experienced dispatchers were already mentally doing something close to what an AI might do, but they were constrained by having to input decisions into a system designed for manual route planning. The AI they eventually built was not replacing human expertise. It was amplifying it by removing the tedious data entry that slowed everything down.

They identified a skeptic and brought them inside the tent.

Every organization has someone who thinks the AI initiative is overhyped nonsense. Usually someone with deep operational knowledge and strong opinions. The failed projects treat skeptics as obstacles to be overcome or ignored. Successful projects find the smartest skeptic and make them part of the core team.

Why? Because skeptics ask questions everyone else is afraid to voice. They poke holes in assumptions. They demand proof that the approach makes sense. And critically, when a well-informed skeptic becomes convinced that something is working, everyone else notices.

A hospital in Toronto made their most skeptical ER physician the medical lead on their patient flow optimization project. He spent the first month explaining in detail why it would not work and why AI could not possibly understand the complexity of emergency medicine. By month three, he had become the system's strongest advocate, not because he had been convinced by presentations but because he had watched it actually improve patient flow during three consecutive busy weekends. His endorsement mattered more than any executive memo.

They picked success metrics that would hurt to miss.

Most AI projects have goals that sound impressive but do not create real accountability. "Improve customer satisfaction." "Increase operational efficiency." "Enhance decision-making quality." These are directions, not destinations.

Successful projects commit to metrics specifically enough that failure would be obvious and embarrassing. "Reduce customer call handle by 90 seconds by the end of Q3." "Decrease prediction error rates below 8% by June." "Enable 40% of loan applications to process without human review by September."

There are common traits about those metrics: they are time-bound, measurable, and tied to specific business outcomes. If you miss them, everyone knows. That specificity creates focus. It forces hard conversations early about whether you are making real progress or just making excuses.

They built feedback loops that work at operational speed.

In successful AI projects, the people using the system can influence how it behaves. Not through formal change request processes that take weeks, but through mechanisms that operate in days or hours.

A manufacturing company in Bangalore built a quality control vision system and created a simple interface that let line workers flag when the AI got something wrong. Not just report errors but explain why the classification was incorrect in ways the system could learn from. Within two months, the accuracy had improved from 73% to 94%, not because the initial model was bad but because the people doing quality control were teaching it their expertise.

They celebrated learning as much as wins.

The first 90 days of any AI project involve a lot of things not working the way you expected. Data is not where you thought it was. Models perform differently in production than in testing. Users interact with the system in ways you did not anticipate.

Failed projects treat these discoveries as setbacks. Successful projects treat them as progress. They are learning what is really true about their situation, which is infinitely more valuable than maintaining comfortable illusions.

When that insurance company in Melbourne discovered that their claims data was structured completely differently than they had assumed, they did not scramble to hide it from executives. They added two weeks to the timeline, documented what they had learned, and adjusted the approach. The CEO's response: "Good. Better to find out now than in six months' time."

EXECUTIVE SPONSORSHIP: BEYOND THE BUDGET SIGNATURE

Here is a pattern you will see in every failed AI initiative: strong executive sponsorship at the beginning, declining engagement over time, sudden reappearance when things go wrong. The executive champion approves the budget, attends the kickoff meeting, shows up for a quarterly review. Then he gets busy with other priorities assuming the team has everything under control.

The project stalls, goes sideways or delivers something that technically works but does not actually solve the business problem. The executive reenters the scene, usually frustrated, trying to figure out what went wrong and who is accountable.

Successful AI transformations have executive sponsors who operate completely differently. They are not just approvers or cheerleaders. They are active participants who do three specific things the missing executives do not do:

They protect the project from organizational antibodies.

Every organization has immune system responses that kill new initiatives. Budget reallocations. Competing priorities. Key team members are getting pulled to "more urgent" work. Policy requirements that made sense for traditional IT but strangle innovation.

A manufacturing VP in Munich watched her AI initiative start to drown in compliance requirements designed for enterprise software deployments. Weekly security reviews, architecture approval boards and change advisory committees. None of it wrong, exactly, but collectively creating a process that would have turned a six-month pilot into an eighteen-month ordeal.

She did not fight the compliance requirements. She created a protected space. They designated the AI initiative as an experimental project with different governances and committed to bringing it into standard processes once they'd proven the value. The VP took personal accountability for the risk. The project moved fast enough to demonstrate value, which made it much easier to navigate compliance later with concrete results to show.

They make themselves available for the challenging decisions.

AI projects generate questions that only executives can answer. Should we integrate with the legacy system or build parallel? Do we prioritize accuracy or explainability? If the

AI identifies problems in how we are currently measuring performance, do we change the AI or change the metrics?

These decisions do not belong to project steering committee agendas. They need executive judgment, and they need it quickly. The successful executives don't just say "I'm available if you need me." They schedule regular office hours. They show up on Slack channels. They make it clear that asking for their input is expected, not an imposition.

They translate between worlds.

There is a language barrier in AI projects. Data scientists speak in precision and probabilities. Business leaders speak about outcomes and impacts. Operations people speak in workflows and exceptions. Everyone is saying important things, but they are often talking past each other.

The executive sponsor's most valuable role is translation. Taking what the data scientist means by "the model has 87% accuracy with a precision of 0.91" and explaining to the business leader that "9 out of 10 times it flags something as high priority, it's right, but it misses about 13% of the cases we'd want it to catch." Taking what the operations manager means by "this doesn't fit our workflow" and helping the technical team understand which parts of the workflow are essential and which could change.

GOING SLOW TO GO FAST: THE PATIENCE PARADOX

The successful 20% consistently do something that looks wrong to outside observers: they move deliberately when everyone else is rushing. While their competitors are racing to deploy AI across the organization, they are running small experiments. While others are signing massive vendor contracts, they are working with scrappy minimum viable products. While everyone else is announcing their AI strategy to the market, they are quietly cleaning their data.

This is not caution. It is not risking aversion or lack of ambition. It is recognition that AI transformation requires learning, and learning requires time.

A pharmaceutical company spent six months on data preparation before building any models. Colleagues thought they were wasting time. When they finally started building, their models worked immediately because the data was clean and well-structured. They saved twelve months compared to competitors who rushed into modeling and then spent a year debugging data problems.

THE ITERATIVE LEARNING APPROACH: FAIL INFORMATIVELY

There is a crucial difference between failure that generates learning and failure that just burns money. The successful 20% also fail constantly. They just fail fast and recover quickly. They learn and move on.

The framework they follow, whether they articulate it this way or not, looks like this:

Pilot: Pick the smallest possible version of the problem you can solve. Not the whole customer service operation, but the 30% of inquiries that follow predictable patterns. Not all preventative maintenance, but the three machines that fail most often. Not the entire claims process, but the straightforward cases that should not require adjustment at all.

Build something minimal. Focus on learning, not perfection. Accept that your first version will be wrong in ways you cannot anticipate.

Learn: Measure what actually happens, not what you hoped would happen. Talk to the people using the system. Ask what works and what does not. Look for surprises, where did reality differ from assumptions?

A healthcare startup in Boston built an AI system to help nurses identify which patients needed additional monitoring. The first version was terrible. It flagged too many patients, generating alert fatigue. When they interviewed the nurses, they learned something fascinating: the AI was not worse than nurses at identifying risk, it was just using different signals. Nurses noticed behavioral cues that were not in the data. The AI noticed physiological patterns too subtle for humans to spot.

They did not declare failure. They declared learning. The AI was seeing something real. The question wasn't whether to use it, but how to combine its strengths with human judgment.

Adjust: Change the approach based on what you have learned. Sometimes that means tweaking the model. Sometimes it means restructuring the data. Sometimes it means completely reimagining the problem.

That healthcare startup rebuilt their system to work differently. Instead of generating alerts, it created a daily list of patients ranked by risk. Nurses used it like a triage tool, focusing their limited attention on patients, the AI thought was highest risk. Accuracy did not change, but utility increased dramatically.

Scale: Only scale what is proven to work. This sounds obvious, but organizations violate it constantly. They pilot something, get mediocre results, and scale it anyway because they have already committed the budget or made promises to stakeholders.

The successful ones are ruthless about this. If it does not work in the pilot, they kill it or fundamentally redesign it. You cannot scale something that does not work.

A European logistics company ran pilots of route optimization in three different regions. One region saw a 12% efficiency improvement. One saw 6%. One saw none. Instead of scaling to all regions, they studied the differences. Turned out the region with no improvement had far more unpredictable delivery windows because of urban traffic patterns. The algorithm could not account for that variability.

They scaled to the regions where it worked. They did not waste money, forcing it to work everywhere. They learned something valuable: route optimization AI works best in environments with predictable constraints. That insight shaped every subsequent AI initiative.

THE DATA THAT DISTINGUISHES WINNERS FROM LOSERS

Researchers at Stanford collected data from 400 AI implementations across different industries and companies of sizes. They were looking for quantitative patterns that separated successful projects from failures. Not opinions about what worked, but measurable differences.

They found six patterns that appeared consistently:

Speed to first production deployment: Successful projects had something running in production within 120 days. Failed projects averaged 240 days before deployment. The extra time was not making projects better. It was creating distance between the problem and the solution, allowing requirements to drift and stakeholders to lose interest.

Ratio of pilot projects to scaled solutions: Failed organizations had 8:1 ratio. Eight pilots for every one thing that scaled.
Successful organizations had 3:1 ratio.
They were either better at picking pilots that would work, or better at killing pilots that wouldn't, or both.

Percentage of initial team members still involved at month six: Successful projects retained 80% of their starting team. Failed projects had lost half their people by month six. Continuity mattered. The team's accumulated learning mattered. Starting over with new people meant relearning lessons.

Executive sponsor meeting frequency: Successful projects had executive sponsors actively engaged weekly. Failed projects had monthly or quarterly engagement. The difference was not micromanagement versus autonomy. It was signal versus noise. Weekly engagement meant problems got addressed when they were small. Monthly engagement meant problems were festered.

Number of end-users involved in design: Successful projects had 10+ people who would eventually use the system involved in shaping it. Failed projects had 2-3 people, usually analysts or managers. The gap between what designers thought users needed and what users actually needed was huge.

Time from "technical success" to "business value": This was the most revealing pattern. Successful projects delivered measurable business value within 30 days of technical deployment. The model worked, and immediately the organization saw impact. Failed projects had this bizarre gap. The model worked technically but business value did not materialize for months or never.

That gap revealed a fundamental problem. They had built something technically impressive that didn't actually solve a business problem.

DOES YOUR APPROACH MATCH THE SUCCESS PATTERN?

Here is the uncomfortable question: How many of these success patterns describe your AI initiative?

If you are planning to start an AI transformation, you can use these patterns diagnostically. If you're already in the middle of one, there are early warning signs.

Let me make this concrete. Answer these questions honestly:

Strategic Clarity

Can you describe your AI initiative's goal in one sentence that a frontline employee would understand? If you stopped the project today, could you articulate what specific business value you would lose?

Executive Engagement

Is your executive sponsor involved weekly? Do they make themselves available for decisions that need their judgment? Have they actively protected the project from organizational obstacles?

Operational Integration

Have you involved the people who will actually use the system in designing it? Have you identified your smartest skeptic and brought them into the project? Do you have feedback mechanisms that work at operational speed?

Learning Orientation

Have you committed to specific, measurable outcomes with dates? Are you treating surprises as learning opportunities? Are you willing to kill initiatives that are not working even if you have already invested in them?

Pace and Scope

Are you trying to prove value in weeks or months, not years? Is your first deployment focused on a narrow, well-defined problem? Are you planning to iterate based on what you learn rather than executing a fixed plan?

If you can honestly answer "yes" to most of these, you are probably in the successful 20%. If you're answering "no" or "sort of" or "we're planning to," you need to stop and address these fundamentals before you go further. That insurance CEO in Melbourne, the one who paused his project six weeks in. In the post project debrief he had later admitted that the hardest part was not admitting that they had started wrong but convincing everyone involved that pausing to course correct was not the same as failing. In projects of this nature, we have been so conditioned to think that momentum means progress, even when you are moving in the wrong direction.

He was right. Momentum feels like progress. Activity feels like achievement. If you are moving fast in the wrong direction, the best thing you can do is stop.

WHAT SUCCESS ACTUALLY LOOKS LIKE

I want to be noticeably clear about something: successful AI transformation does not look like the vendor demos or conference presentations. It looks messier. More gradual. Less revolutionary.

Success looks like a logistics company that reduced delivery costs by 11% over eight months through better route optimization. Not 40%. Not overnight. Real money saved through AI that actually works.

Success looks like a hospital that decreased patient wait times in their imaging department by 23 minutes on average because AI scheduling understood patterns human schedulers missed. Not a complete transformation of healthcare. Just one department works noticeably better.

Success looks like a manufacturing operation that reduced defect rates from 2.1% to 1.4% because computer vision caught problems human inspectors were missing. Not perfect quality. Just measurably better quality.

These do not make headlines. They do not win innovation awards. However, they generate real value. They solve real problems. They make organizations tangibly better at what they do.

Here is what makes them sustainable. The people in those organizations understand why AI works and how to improve it. They are not dependent on vendors or consultants to keep the system running. They have built capability, not just deployed technology.

That is the pattern worth pursuing. Not dramatic transformation promised in vendor pitches, but genuine improvement achieved through thoughtful implementation.

The successful 20% are not doing anything magical. They are just avoiding the mistakes everyone else makes. They start with problems instead of solutions. They are involving users instead of imposing systems. They are measuring what matters instead of what's easy to measure. They are learning from failures instead of hiding them.

That is the pattern. That is what success looks like when you strip away the hype and look at what really works.

The next section of this book shows you exactly how to implement this pattern in your organization. Not theory. Not aspiration. Specific frameworks and tools you can use starting tomorrow.

First, you need to know where you are starting from. The right approach depends entirely on your current AI maturity level, and most organizations dramatically overestimate how ready they actually are.

Let us figure out where you really are.

CHAPTER 5

THE AI MATURITY MODEL

The CTO of a mid-sized financial services firm in Dublin printed out a survey. One of those vendor assessment tools that promise to evaluate your organization's AI readiness in fifteen minutes. She'd filled it out during a lunch break, feeling rather good about their responses. The results came back: "AI Ready - Proceed with Confidence."

Three months later, their first AI initiative was drowning. Models that worked beautifully in testing failed spectacularly in production. The data they thought was clean turned out to be riddled with inconsistencies. The team they had assembled did not have the skills they actually needed. Business users did not trust the output.

The survey asked if they had data which they had plenty of. It asked if they had executive support. The CEO approved the budget personally. It asked if they had technical expertise to which the answer was, they had a strong IT department. Every question they could answer yes but none of those questions mattered.

Having data is not the same as having usable data. Having executive support is not the same as having executive understanding. Having technical expertise in traditional IT is not the same as having expertise in AI implementation.

Most AI readiness assessments are useless because they measure the wrong things. They measure inputs and intentions rather than capabilities and conditions. They tell you whether you have checked certain boxes, not whether you are actually prepared to succeed.

This chapter introduces a different kind of assessment. Not a survey you can game, but a framework for honestly evaluating where your organization stands and what that means for how you should approach AI transformation.

THE FIVE LEVELS OF AI MATURITY

Organizations do not exist at a single point of AI readiness. They exist across a spectrum, and understanding where you fall on that spectrum determines everything about how you should proceed.

I have studied many organizations attempting AI transformation, and they cluster into five distinct maturity levels. Each level has characteristic strengths, typical blind spots, and specific strategies that work. Try to implement a Level 4 strategy when you are actually at Level 2, and you'll fail. Stay stuck at Level 1 when you're ready for Level 3, and you're leaving value on the table.

HERE IS THE FRAMEWORK

Level 1: AI Unaware

Organizations at this level are not actively thinking about AI. It is not that they are against it or don't believe in it. It's just not on their radar. They are solving business problems with traditional approaches, and those approaches are mostly working well enough so that AI does not feel urgent.

These aren't necessarily small or unsophisticated organizations. Some are phenomenally successful in their markets.

They have just reached a stable equilibrium that does not currently require AI capabilities.

Level 2: AI Exploring

Organizations here are curious about AI. They are reading articles, attending conferences, and watching what competitors do. Leadership is asking questions about whether they should be "doing something with AI." There might be informal experimentation happening, someone in IT tinkering with machine learning tutorials, a motivated manager exploring chatbot possibilities.

However, there is no coherent strategy. No budget. No real commitment. Just exploration and increasing awareness that AI might eventually matter to their business.

Level 3: AI Implementing

These organizations have moved beyond exploration into actual implementation. They have one or more AI pilots running. They have allocated budget. They have formed teams. They are collaborating with vendors or building capabilities.

This is the dangerous level. You are committed enough to spend real money but not yet mature enough to spend it wisely. Most organizations that ended up in failure graveyard were at Level 3 when they stumbled.

Level 4: AI Scaling

Organizations at this level have proven that AI can deliver value through successful pilots. Now they are figuring out how to scale what works. They are moving from project mode to program mode. They are building repeatable processes. They are developing internal expertise. They are integrating AI into standard operations rather than treating it as a special project.

Success here requires different skills than success at Level 3. You are no longer just proving concepts work. You are benefiting from them.

Level 5: AI Transforming

These rare organizations have embedded AI so deeply into their operations that it is no longer a separate initiative. AI is just how they work. They are using AI to reimagine business models, not just optimize existing processes. They are building competitive advantages that it would be impossible to replicate without similar AI capabilities.

Very few organizations reach Level 5 and reaching it is not necessary for success. Understanding that Level 5 exists helps clarify what mature AI transformation actually looks like.

Level 1: AI Unaware – When Ignorance Might Be Rational

Being at Level 1 is not failure. For some organizations, in some circumstances, not pursuing AI is the right strategic choice.

A small law firm in Vancouver specializing in estate planning might be at Level 1. They have fifteen lawyers, established processes that work, clients who value personal relationships. They are profitable. Their competitive advantage comes from expertise and reputation, not operational efficiency.

Should they be exploring AI? Eventually. But if they are not, that is obviously not wrong. The opportunity cost of time spent learning about AI versus time spent serving clients might genuinely favor clients.

The question is not "Should every organization pursue AI?" The question is "Does remaining at Level 1 expose us to risks we should address?"

Those risks are real. Markets change. Disruptors emerge. Competitors evolve. Client expectations shift. The law firm that is fine without AI today might find in three years that clients expect instant document generation, automated compliance checking, and research assistance that their manual processes cannot match economically.

Characteristic Strengths of Level 1 Organizations:

They are not wasting money on ill-conceived AI initiatives. They are focused on their core business. They have not created change fatigue by chasing every innovative technology trend.

Typical Blind spots:

They may not recognize when their competitive environment is shifting. They may underestimate how quickly AI capabilities are improving and how accessible they're becoming. They may be vulnerable to disruption from competitors who leverage AI or digital transformation more effectively.

Right Strategy for Level 1:

Stay aware without panicking. Set up lightweight monitoring: have someone track AI developments in your industry, attend occasional conferences, maintain network connections who can alert you to shifts that matter. Do not launch initiatives just to say you are "doing AI," but do not ignore signals that your market is changing.

The trigger to move from Level 1 to Level 2 is not that AI exists. It is that you are encountering specific business problems where AI might offer meaningfully better solutions than your current approaches, or that competitors are using AI in ways that threaten your position.

Level 2: AI Exploring – Curiosity Without Commitment

Most organizations currently sit at Level 2. They are aware of AI. Leadership mentions it in strategy meetings. There might be a task force studying possibilities. Nothing formally is happening in terms of real implementation.

A manufacturing company makes specialized industrial components. Their CEO read an article about predictive maintenance. Their operations director attended a conference where someone presented a compelling case study about quality control AI. They are intrigued. They have asked IT to "look into what we should be doing with AI."

This is Level 2. Curiosity without commitment. Awareness without action.

The danger here is getting stuck. Exploration can become a comfortable steady state. You are aware enough to feel like you are not ignoring AI but not committed enough to actually learn anything real about whether it can help your business.

Characteristic Strengths of Level 2 Organizations: They are building awareness without rushing into expensive mistakes. They have time to learn from other organizations' successes and failures. They can be strategic about where to focus when they do commit.

Typical Blind spots:

They often confuse learning about AI in general with understanding how AI might specifically help their business. They may be waiting for "the right time" to start, not recognizing that the right time is usually "before it feels urgent."

They tend to delegate AI exploration to staff who, while capable, do not have authority to actually make things happen. The result is reports and presentations that go nowhere.

Right Strategy for Level 2:

Set a deadline for moving from exploration to decision. Not "we'll implement AI by Q3," but "by Q3, we'll decide whether to commit to an AI pilot or explicitly decide AI isn't a priority for the next 18 months."

Create a small team, maybe three people, with a specific mandate: identify the three business problems where AI might deliver ten times better solutions than current approaches. Not a comprehensive AI strategy. Just three specific problems worth investigating.

Give them budget for the investigation itself. Let them hire consultants if they need external expertise. Let them visit organizations that have succeeded with similar applications. Let them run exceedingly small tests if needed to validate that the approach makes sense in your context.

Then decide. If the investigation reveals compelling opportunities, commit. If it doesn't, park AI for now and revisit in a year. Either way, you are learning and progressing rather than perpetually exploring.

Level 3: AI Implementing – Where Most Failures Happen

Level 3 is where organizations have committed. They have one or more AI initiatives underway. They have allocated real budget. They have formed teams or hired consultants. They are building things.

This is also where 80% of failures occur.

Why? Because at Level 3, you are trying to do something genuinely new to your organization, but you do not yet know what you don't know. You have enough commitment to spend significant money, but not enough experience to spend it wisely. You are confident enough to start but not experienced enough to recognize when you are heading toward failure.

A retail chain in Cape Town decided to implement personalized marketing using AI. They are at Level 3. They hired a data science team. They invested in a marketing automation platform. They allocated six months and a substantial budget. Everyone is committed and enthusiastic.

Four months in, they are struggling. The customer data they thought was clean needs extensive cleaning. The models generate recommendations, but the marketing team doesn't trust them enough to actually use them. The system is technically functional but does not deliver business value.

This is the classic Level 3 failure pattern. Not because of incompetence or bad intentions, but because they did not yet understand what successful AI implementation actually requires.

Characteristic Strengths of Level 3 Organizations: They have overcome organizational inertia. They are actually doing something rather than just talking about it. They are building experience and learning, even if painfully. They have committed enough resources that success is possible.

Typical Blind spots:

They often confuse activity with progress. They may not recognize that technical success and business value are different things. They frequently underestimate how much change management matters relative to technical implementation.

They tend to believe that because they have hired smart people and allocated budget, success will naturally follow. They do not realize that success at Level 3 requires intense, specific focus on bridging the gap between technical capability and operational reality.

Right Strategy for Level 3:

Relentlessly focus on one question: "Are we generating measurable business value yet?" Not "Is the model working?" Not "Are we making progress?" Are we actually solving a business problem in a way that creates value?

If yes, great. Scale it. Learn from it. Apply those lessons to the next initiative.

If not, figure out why with brutal honesty. Is it technical? Is it adoption? Is it that you are solving the wrong problem? Do not fall into the trap of assuming more time will fix things. More time fixes some problems and entrenches others.

Set kill criteria before you start. "If we don't see X improvement in Y metric by Z date, we stop or fundamentally pivot." Then actually follow those criteria. Most Level 3 organizations set these criteria and then ignore them when the deadline arrives.

Get outside perspective. Someone who has successfully implemented AI in a similar context. Someone who can tell you whether what you're experiencing is normal growing pains or signs of fundamental problems. Not a vendor trying to sell you something. Someone with no stake in your success or failure beyond professional reputation.

The goal at Level 3 is not to implement a lot of AI. It's to successfully implement one or two initiatives well enough that you understand what success actually requires. You are paying your tuition fees to learn how AI works in your specific organizational context.

Level 4: AI Scaling – Industrializing What Works

Organizations at Level 4 have proven that AI can work. They have successful pilots that deliver business value. Now they face a different challenge: how to scale what works without losing what made it work.

A logistics company in Singapore successfully implemented route optimization for their premium delivery service. It worked beautifully. Reduced costs, improved delivery times, drivers liked it. Now they want to scale it across their entire operation, which is ten times larger and more complex.

This is the Level 4 challenge. You are no longer proving concepts. You are industrializing them. Different skills required. Different risks encountered.

Characteristic Strengths of Level 4 Organizations: They have proof that AI delivers value in their context.

They have learned from initial implementations. They have momentum and credibility. They have started building internal expertise rather than being purely dependent on vendors.

Typical Blind spots:

They may assume that what worked in a pilot will automatically work on a scale. They might not recognize that scaling requires operational discipline that innovation does not. They can become overconfident, thinking they've mastered AI implementation when they have really just succeeded once.

The transition from Level 3 to Level 4 is dangerous because success can breed carelessness. You stop doing careful, methodical work that made you successful because you think you've figured it out.

Right Strategy for Level 4:

Build repeatable processes. Document of what worked and lessons learnt in your successful pilots. Not just the technical architecture, but the full recipe: how you engaged stakeholders, how you managed change, how you measured success, how you managed problems.

Create an internal AI center of excellence, but make it a service organization, not an ivory tower. Their job is not to control all AI activity. It is to make it easier for various parts of the organization to succeed with AI by sharing learnings, providing templates, and offering expertise.

Develop your people systematically. You have probably been relying on a few key individuals. That does not scale. You need to build AI literacy across the organization and deep expertise in pockets where it matters most.

Start thinking in terms of platforms, not just projects. How can the infrastructure you build for one use case serves multiple use cases? How can the data pipelines you create for one application make other applications easier?

Here is the critical thing about Level 4: you scale what works, not what you hoped would work. If a pilot delivered marginal value, don't scale it hoping it will get better. It probably won't. Kill it and redirect those resources to scaling the things that genuinely succeeded.

Level 5: AI Transforming – When AI Becomes Invisible

Very few organizations reach Level 5, and it is worth questioning whether reaching it should even be a goal. At Level 5, AI is so embedded in operations that it stops being a separate thing you talk about. It is now part of your company's DNA. It is how work gets done.

An e-commerce platform operating across Southeast Asia is at Level 5. AI handles demand forecasting, inventory optimization, dynamic pricing, fraud detection, customer service routing, personalized recommendations, and logistics planning. These are not AI projects. They are just business operations that happen to use AI.

The teams running these systems do not think of themselves as "doing AI." They are solving business problems, and AI is one of their tools. They are as comfortable with AI as they are with databases or networks. It is infrastructure, not innovation.

Characteristic Strengths of Level 5 Organizations: They have built sustainable competitive advantages through AI. They can move quickly on new opportunities because the foundational capabilities are in place. They have developed organizational muscle memory around AI implementation.

Typical Blind spots:

They may become complacent, assuming their current AI capabilities will remain sufficient. They might not recognize when the AI landscape shifts in ways that require rethinking their approach. Success can make you less hungry, less careful, less aware of threats.

Right Strategy for Level 5:

Stay paranoid. Amazon famously operates with a "Day 1"

mentality even though they are decades into their AI transformation. The moment you think you have arrived is the moment you start falling behind.

Keep investing in capability development. Your competitors are trying to reach Level 5. Your advantage is temporary unless you are continuously improving.

Look for opportunities to use AI not just to optimize existing business models but to enable new ones. That is the real power of Level 5, you have the capability to imagine and execute business models that would be impossible without deep AI integration.

THE SELF-ASSESSMENT: WHERE ARE YOU REALLY?

Now the uncomfortable part. Where is your organization actually?

Most organizations overestimate their maturity by at least one level. They think they are at Level 4 when they are really at Level 3. They think they are at Level 3 when they are really at Level 2 with one pilot project.

Here is an honest diagnostic. Answer these questions, and let the answers tell you where you actually are:

Strategic Questions

1. Do you have a written AI strategy that specifically identifies which business problems you are solving and why AI is the right approach? If you have a document that talks about "exploring AI opportunities" or "becoming AI-driven," that is not a strategy. That is aspiration.
2. Can your executive team articulate, without preparation, what specific business value you expect from your AI initiatives and when you expect to see it?

3. Have you committed multi-year investment in AI capabilities, not just individual projects? Is that commitment reflected in actual budget allocation, not just strategic intent?

Execution Questions

4. How many AI initiatives have gone from pilot to production and are currently delivering measurable business value? Not "in production" meaning deployed but unused. In production and delivering value.
5. For initiatives that did not succeed, have you documented why they failed and applied those lessons to subsequent initiatives? Or do you tend to quietly shelve failed projects and move on?
6. Do you have repeatable processes for implementing AI, or does each initiative feel like starting from nothing?

Cultural Questions

7. Business users actively request AI capabilities to solve problems they are encountering? Or is AI something that IT or data science teams push into the organization?
8. When AI systems make recommendations, do operational staff generally trust and act on them? Or do they routinely second-guess or ignore the AI?
9. Is AI literacy widespread across your organization, or concentrated in technical teams? When we talk of Ai Literacy, we mean is there basic understanding of what AI is, what it can and can't do?

Capability Questions

10. Do you have people who understand both AI technology and your specific business domain? Not just data scientists who are learning business, not just business-

people who are learning AI, but people who genuinely operate at the intersection?

11. Can you implement new AI use cases without external consultants? Even if you choose to use consultants, could you do it yourself if necessary?
12. Is your data infrastructure actually ready for AI, or do you discover data quality problems every time you try to use the data?

MATURITY SCORING

If you answered "yes" to fewer than 4 of these 12 questions, you are at Level 2. You're exploring but not yet implementing it in any meaningful way.

If you answered "yes" to 4-7 questions, you are at Level 3. You are implementing, but not yet consistently successful.

If you answered "yes" to 8-10 questions, you are at Level 4. You have proven AI can work and you are scaling what succeeds.

If you answered "yes" to 11-12 questions, you are at Level 5. AI is deeply embedded enough that it is becoming invisible infrastructure.

WHAT YOUR LEVEL MEANS FOR YOUR STRATEGY

Your AI maturity level determines everything about how you should approach AI transformation. The strategy that works at Level 4 will fail catastrophically at Level 2.

If you are at Level 1 or Level 2

Do not start with an ambitious AI transformation program. Start with AI education. Get key leaders to understand what AI is and is not. Visit organizations that have succeeded with AI in contexts like yours. Run small experiments that let you learn without committing significant resources.

Your goal is not to implement AI. Your goal is to develop enough understanding to make an informed decision about whether implementing AI makes sense for your organization.

Period: Depending on size of organization but three months.

Budget: minimal, mostly time and attention from key leaders.

Success metric: an informed decision about whether to proceed to Level 3.

If you are at Level 3

Focus obsessively on successfully implementing one or two small, well-defined initiatives. Not pilot projects that you will evaluate later. Actual implementations that deliver measurable business value.

Resist the temptation to launch multiple initiatives simultaneously. Resist pressure to pursue AI broadly. You are learning how to make AI work in your organizational context. Learning requires focus.

Period: Depending on size of organization but roughly six months per initiative.

Budget: Enough to matter, not enough to threaten the organization if you fail.

Success metric: Measurable business value from at least one AI implementation.

If you are at Level 4

Build the infrastructure, technical and organizational, that lets you scale what works. Create templates, develop processes, train people, establish governance. Think about how to industrialize success rather than just celebrating individual wins.

Start looking for opportunities to use AI capabilities you have built for one purpose to solve other problems. You have paid the learning cost once. Get multiple returns from it.

Period: Depending on size of organization but roughly 12 months.

Budget: Significant investment in platforms and capability development, not just individual projects.

Success metric: Multiple AI implementations delivering value with decreasing time-to-value for each new implementation.

If you are at Level 5

You do not need advice from this book about how to implement AI. You need to stay paranoid, keep investing in capabilities, and look for opportunities to use AI not just to optimize what you do but to reimagine what is possible.

THE UNCOMFORTABLE TRUTH ABOUT MATURITY

Most organizations resist honest self-assessment because the results are uncomfortable. Nobody wants to admit they are at Level 2 when competitors claim to be at Level 4. Nobody wants to tell their board that they need to slow down and build foundations when everyone is speeding up.

Misdiagnosing your maturity level can be extremely expensive. Organizations that implement Level 4 strategies when they are actually at Level 2 don't just fail to get results. They burn money, exhaust teams, and create organizational skepticism that makes future AI initiatives harder.

That CTO in Dublin, the one whose company's "AI Ready" assessment proved wildly optimistic? Six months after their initial failure, they did an honest assessment using this framework. Turned out they were at Level 2, maybe early Level 3 at best. They had tried to implement a Level 4 strategy.

They reset. Scaled back ambitions. Picked one narrow use case. Invested in understanding their data. Brought business users into the design process. Eighteen months later, they had something working. Small, not revolutionary, but genuinely delivering value.

"The hardest thing was admitting we weren't where we thought we were," she noted. "Felt like going backwards, but it was the only way to eventually move forward."

She was right. Honest assessments can feel like backwards movement. It is the foundation for you to build the house on.

You cannot optimize yourself for a race you are not ready to run. You cannot scale what you have not proven works. You cannot transform what you do not understand.

Know where you are. Accept what that means for what you should do next. Execute the right strategy for your actual maturity level.

That is how organizations move from wherever they are to wherever they want to be.

The next chapter assumes you have done that assessment honestly. It provides the strategic foundation needed to move forward successfully, whatever level you are starting from. It requires you to be honest about where you are starting from.

CHAPTER 6

THE STRATEGIC FOUNDATION

Fifteen executives sat around a table listening to the Chief Digital Officer talking about their AI strategy, clicking through slides dense with buzzwords: "leverage," "synergize," "revolutionize," "transform."

A hand went up. The CFO, asked the question that stopped the room cold: "What problem are we actually trying to solve?"

Silence.

The CFO continued, "we've been talking about neural networks and machine learning platforms for fifteen minutes. What specific business problems are we solving or what particular outcomes are we trying to achieve? What metric improves if this all works?"

More silence. Then the CDO did what desperate executives do when caught without substance: He went back on a slide and started reading bullet points.

That meeting, which a participant later described as "the moment we realized we'd been planning our AI strategy backwards," captures something essential about why strategic foundations crumble. Technology had become the strategy. Everyone in that room knew they needed to "do AI" because everyone was doing AI. Nobody could articulate why their specific organization, with its specific challenges and opportunities, needed AI beyond the vague sense that the future required it.

Six months later, that same company would spend millions on a data platform that precisely solved nothing. The CFO had voted against it but was outvoted.

THE STRATEGY VACUUM

Here is what most AI strategies look like in practice: a document that reads suspiciously like the vendor's marketing materials, sprinkled with internal jargon for authenticity. Lots of words like "transformation" and "innovation." Charts showing adoption curves trending upward and to the right, as if that's the only direction lines are legally allowed to travel. Maybe a few use cases, usually the same ones everyone else is implementing because someone saw a competitor announce them.

What these documents lack is the thing strategy requires: hard choices about what you will not do, informed by clear thinking about what you're uniquely positioned to do.

Real strategy hurts. It means saying no to exciting possibilities because they do not serve your specific context. It means admitting your organization is not ready for certain applications, even if they are generating headlines. It means accepting that the most compelling AI use case might not be the one that moves your most critical business metric.

The companies that succeed with AI, the ones in that narrow band of the successful twenty percent, start differently. They do not start with technology at all. They start with business problems worth solving and business opportunities worth pursuing, then ask whether AI might be part of the solution. Sometimes it is. Often it is not, or at least not yet, not in the way initially imagined.

This chapter is about building that foundation. Not the AI foundation, which comes later. The strategic foundation that must exist first before you commit budget or hire data scientists or sign vendor contracts. Get this wrong and everything built on top of it, no matter how technically sophisticated, becomes expensive architecture without purpose.

YOUR AI NORTH STAR: VISION BEYOND PLATITUDES

Every organization needs what I call an AI North Star; a clear articulation of what success looks like that is specific enough to guide decisions and honest enough to be useful. Not "become an AI-driven organization," which means nothing. Not "leverage AI to enhance operational efficiency," which could mean anything. Something real.

A regional healthcare provider I worked with in Melbourne articulated theirs this way: "Reduce administrative burden on clinical staff by forty percent within two years, freeing time for patient care, while maintaining or improving documentation quality and compliance."

Notice what that does. It names a specific problem: administrative burden. It quantifies the target: forty percent reduction. It sets a period: two years. It acknowledges constraints: quality and compliance cannot suffer. And critically, it connects the AI initiative to a core organizational value: more time for patient care.

That North Star made a hundred subsequent decisions easier. When a vendor pitched an impressive diagnostic AI system, the executive team could evaluate it against their stated priority. Might it be valuable someday? Sure. Was it the administrative burden problem they had committed to solving first? No. Pass.

Your North Star needs three components, and all three need to be honest.

The Business Outcome: What specific result are you pursuing? Revenue increase? Cost reduction? Customer satisfaction improvement? Risk mitigation? Pick the thing that matters most to your organization right now, not the thing that sounds most impressive in a presentation. One manufacturer of industrial equipment landed on "reduce warranty claims by identifying product failures before they reach customers." Unglamorous. Measurable. Connected to a bleeding wound in their P&L.

The Strategic Rationale: Why this outcome, why now, why AI? This is where you articulate the business logic that makes this investment make sense. Maybe your customer demographics are shifting, and your traditional approaches no longer work. Maybe competitors are doing something that is changing customer expectations. You're drowning in data you cannot analyze with current tools. Whatever it is, name it clearly.

A logistics company in Frankfurt explained their rationale this way: "E-commerce growth has increased our delivery volume by two hundred percent in three years, but our route optimization approaches were designed for predictable B2B patterns. We are now unprofitable with thirty percent of deliveries because we cannot plan efficiently for high-variety, low-predictability residential routes. Traditional optimization software cannot manage this complexity at our scale. We need dynamic, learning systems that improve as patterns shift."

That is strategic rationale. Specific. Rooted in business reality. Explaining not just what they want to accomplish but why traditional approaches will not work.

The Success Measures: How will you know if you are succeeding, and how will you know if you are failing? This requires metrics you can actually measure, baselines you actually know, and targets that are ambitious but achievable. Not "improve customer experience" but "increase Net Promoter Score from 42 to 55" or "reduce customer service resolution time from 48 hours to 6 hours" or "decrease customer churn from 23% annually to under 15%."

One financial services firm set their success measure as "process loan applications in under four hours with ninety-five percent automation and lower default rates than our current manual review process." That is testable. You will know whether you hit it. And notice the sophistication: they didn't just optimize for speed; they insisted default rates could not increase. They understood that going fast in the wrong direction is worse than going slow in the right one.

THE BUSINESS-FIRST AI CANVAS

Strategy needs scaffolding. Something to structure thinking, force challenging questions and expose gaps in logic before they become gaps in implementation. What I use with organizations is a framework I call the Business-First AI Canvas, which we work through in a series of focused sessions before anything gets built or bought.

It has organized around seven questions, each designed to prevent a specific category of strategic failure.

Question One: What business problems are we solving, and how do we know it is real?

This sounds obvious until you watch organizations struggle with it. The temptation is to answer with solutions disguised as problems: "We need better data analytics" or "We should be using machine learning." Those are not problems; they are proposed solutions. Keep pushing until you get to actual business pain.

A retail chain had initially identified their problem as needing to personalize customer communications. After some questioning by the board, what emerged was that email open rates had fallen from 28% to 11% over three years, and conversion rate from email to purchase had dropped by half. Customers told them in surveys that their communications felt generic and irrelevant. They were losing revenue and customer engagement, and the current segmentation approach of grouping customers into five broad categories was not working anymore.

That is a real problem. Quantified. Connected to business outcomes. And notice something important: AI might help solve it, but so might better copywriting, or different email timing, or fundamental changes to what they're offering. The problem statement does not prejudge the solution.

How do you know it is real? Evidence. Customer complaints. Revenue data. Operational metrics. Employee surveys. Competitive losses you can document. Not hunches, not boardroom conventional wisdom, not "everyone knows this is an issue." If you do not have data proving the problem exists, your first investment should not be in AI to solve it, it should be in measurement systems to understand whether it actually needs solving.

Question Two: What is the current cost of this problem?

If you cannot quantify what the problem costs you now, you will never be able to build a credible business case for investing in solutions. You need to think about cost broadly: direct financial cost, opportunity cost, strategic cost, and cultural cost.

Direct financial cost is the easiest. A manufacturing operation calculated that unplanned equipment downtime was costing them $450,000 monthly in lost production, overtime labor, and rush shipping to meet customer commitments. That's a number with which you can work. Any solution that costs less than $450,000 monthly and prevents that downtime has a positive ROI, assuming it actually works.

Opportunity cost is harder to quantify but often larger. What are you not doing because resources are consumed managing this problem? A software company determined that their engineering team spent thirty percent of their time debugging and fixing issues that could have been caught earlier in development. The cost wasn't just the salary of engineers doing reactive work, it was the new features and products they weren't building because they were fixing what broke. They calculated the opportunity cost at roughly $8 million annually in deferred revenue from delayed product releases.

Strategic cost asks: what does this problem do to our competitive position? A logistics firm realized that their inability to provide accurate delivery time predictions was costing them bids for contracts with major e-commerce companies who had made precise delivery windows a requirement. They were not just losing money on inefficiency; they were being excluded from their fastest-growing market segment.

Cultural cost is hardest to quantify but can be most damaging. When employees spend their days on repetitive work that feels pointless, you pay in turnover, in reduced productivity, in the slow erosion of morale that makes good people start updating their LinkedIn profiles. One insurance company calculated that their claims processing staff had sixty-eight percent annual turnover, and the average employee lasted only eighteen months before quitting. Recruiting and training costs alone were substantial, but the actual cost was in institutional knowledge walking out the door and in the constant presence of inexperienced staff making expensive mistakes.

Question Three: What would success look like, specifically?

This is where vague aspirations go to die. "Improved efficiency" is not success, it is a direction. Success needs numbers, timelines, and consequences.

Work toward metrics you can measure objectively. A hospital system defined success for their diagnostic support AI as: "Reduce average time from symptom presentation to accurate diagnosis by thirty percent, measured through EHR timestamp analysis, while maintaining or improving diagnostic accuracy rates compared to current baseline of 87% first-diagnosis accuracy, to be achieved within eighteen months of implementation."

That is testable. It has numbers. It has a timeline. It acknowledges baseline performance and it includes a critical constraint: do not improve speed at the cost of accuracy. These constraints matter as much as the targets because they prevent optimization in one dimension that destroys value in another.

Sometimes success includes qualitative elements that matter, and that is fine if you can operationalize them. "Improved doctor satisfaction" is too vague, but "increase physician satisfaction with diagnostic support tools from current survey score of 5.2 to above 7.5 on a 10-point scale, with at least seventy percent adoption rate among clinical staff" gives you something measurable.

Question Four: What would failure look like, and how will we know if we are heading there?

Organizations are usually better at defining success than failure, which is backwards. You need to know what failure looks like so you can detect it early and kill initiatives that are not working before they metastasize into money pits.

Define failure with the same specificity as success. A financial services company set failure criteria for their fraud detection AI: "If false positive rates exceed twenty percent, or if fraud detection rates don't improve beyond current ninety-two percent baseline within six months of implementation, or if processing time per transaction increases beyond 1.2 seconds, the initiative is considered failed and we revert to enhanced traditional rule-based systems."

Notice what that it does: it gives them objective triggers for cutting losses. No endless meetings debating whether to continue. No sunk cost fallacy keeping alive something that is not working. The metrics tell you and they have built in multiple failure modes, because projects rarely fail in only one dimension.

Early warning indicators matter too. What signals suggest you are on a path toward failure before you arrive there? Declining user adoption rates, increasing complaints about accuracy and growing workarounds where people bypass the system. Additionally technical debt accumulating faster than capability improvement or budget overruns that follow a pattern of consistent underestimation. Team turnover accelerating is also a key warning. These are leading indicators. By the time you hit your defined failure criteria, you have already lost time and money, you cannot recover. The warning signs let you intervene earlier.

Question Five: What resources, capabilities, and constraints do we have?

Most organizations dramatically overestimate their readiness for AI implementation. They imagine resources and capabilities they do not have, and they ignore constraints that will absolutely matter.

Start with resources. Budget is obvious but looks beyond initial implementation costs to ongoing operational costs. AI systems need feeding, maintaining, and improving. Models need retraining. Data pipelines need monitoring and infrastructure needs scaling. A telecommunications company budgeted $3 million for implementation of a customer service AI but allocated only $200,000 annually for maintenance and improvement. Within two years, the system had degraded so badly from lack of maintenance that they had to rebuild it, essentially paying for implementation twice.

Capabilities are often the biggest gap. Do you have people who can evaluate AI vendors effectively? Do you have data scientists or engineers who can implement solutions? Do you have IT infrastructure that can support AI workloads? Do you have data in usable form?

Constraints are what you bump into when optimistic planning meets reality. Regulatory constraints in healthcare or financial services limit what data you can use and how. Technical constraints of legacy systems that can't integrate with modern AI tools. Political constraints where powerful stakeholders can block initiatives. Budget constraints that are real, not aspirational. Timeline constraints where business needs demand faster results than realistic implementation allows.

> One retail bank acknowledged a constraint that shaped their entire strategy: "We cannot, under any circumstances, risk an incident where customer data is exposed or misused. Our regulatory environment and reputation make this an absolute red line. This means we cannot use any AI approach that we don't understand completely, that requires external data sharing, or that puts customer information in cloud environments we don't control." That constraint eliminated certain vendor solutions and certain approaches, which was fine. It is better to acknowledge constraints upfront than discover them after you have invested in solutions, they make impossible.

Question Six: What is our implementation philosophy?

How you approach implementation matters as much as what you implement. Some organizations benefit from bold, transformative leaps. Most benefit from careful, iterative, learning-based approaches. Know which you are.

The careful, iterative path looks like this: Start small. Prove value in a contained environment. Learn from what works and what does not. Adjust based on evidence. Scale what succeeds. Kill what fails. This approach takes longer but accumulates less risk and generates organizational learning that makes subsequent initiatives more likely to succeed.

> A manufacturing company articulated their philosophy: "We will pilot every AI initiative in a single facility where failure won't threaten business continuity. We will run pilots for at least three months, measuring defined success criteria weekly. Only pilots that meet seventy-five percent of success criteria proceed to broader implementation. Anything that fails gets analyzed for lessons learned before we try again."

The bold, transformative approach works when you have burning platforms, when competitive threats demand fast movement, when organizational culture embraces risk, and when leadership can withstand inevitable failures along the way. A startup in the logistics space took this approach because they had to: "Our traditional competitors are twenty times our size with better customer relationships. Our only chance is to be fundamentally better in delivery experience, which means building our entire operation around AI-optimized routing and prediction from day one. We do not have the luxury of careful pilots. We build it right or we fail."

Both philosophies can work. What does not work is inconsistency, where you say you are being careful but then make impulsive, high-risk bets. Or saying you are bold but then getting paralyzed by normal setbacks. Pick an approach that matches your organization's reality and tolerance for risk, then commit to it.

Question Seven: How does this AI initiative connect to broader business strategy?

AI cannot be an isolated initiative living in its own strategic bubble. It must serve your core business strategy, advancing objectives you have already committed to pursuing. If it does not, it's a distraction regardless of how technically impressive it might be.

Map the connection explicitly.

> A logistics company's business strategy was "become the premium provider for time-sensitive deliveries, commanding a thirty percent price premium over standard services by delivering unfailing reliability." Their AI initiative connected directly: "Use predictive models for route optimization and delay prevention to achieve 99.2% on-time delivery rates, creating the reliability that justifies premium pricing."

That is alignment. The AI serves the business strategy. If AI does not improve reliability, it's not valuable regardless of how sophisticated the technology might be. That clarity makes prioritization easy and prevents scope creep into interesting but strategically irrelevant capabilities.

Sometimes the honest answer to "how does this connect to broader strategy" is "it doesn't, really, we just think AI seems important." That answer should kill the initiative. You do not have unlimited resources or attention. Spend both on things that matter to your strategy.

THE VALUE-FEASIBILITY MATRIX: FINDING YOUR STARTING POINT

Strategy needs prioritization. You probably have more potential AI applications than you have resources to pursue. The Value-Feasibility Matrix helps you figure out where to start, matching ambition to capability in ways that let you build momentum.

Draw two axes. Vertical axis: Business Value, ranging from low to high. Horizontal axis: Technical Feasibility, ranging from low to high. Now plot every potential AI use case you are considering onto this matrix based on honest assessment of both dimensions.

Business value asks: If we implement this successfully, how much does it move metrics we care about? A use case that might save $100,000 annually has lower business value than one that might save $5 million annually or open a $20 million market opportunity. Be realistic about value, accounting for both the magnitude of improvement and the probability of achieving it.

Technical feasibility asks: Given our current capabilities, resources, and constraints, how difficult is this to be implemented successfully? Feasibility considers data availability, technical complexity, integration requirements, organizational readiness, and the dozen other factors that determine whether implementation is straightforward or nightmarish.

The matrix creates four quadrants, each suggesting different strategies.

High Value, High Feasibility: These are your "Quick Wins," the initiatives that deliver significant business value and are realistically achievable with your current capabilities. Start here. These build momentum, generate credibility, create organizational learning, and fund subsequent initiatives. If you do not have anything in this quadrant, your strategy is probably too ambitious for your current state.

A regional insurance company identified "automate routine claims processing for claims under $5,000 with clear documentation" as a Quick Win. This was high value because these claims represented forty percent of their processing volume and consumed enormous staff time. It was high feasibility because the data was clean, the business rules were clear, and the infrastructure was adequate. They implemented it in four months, reduced processing time by seventy percent, and freed up staff for complex claims requiring human judgment. That success gave them credibility to pursue more ambitious initiatives.

High Value, Low Feasibility: These are your "Strategic Investments," initiatives that would deliver significant value but require substantial capability first. These are your medium-term objectives. You do not start here, but you invest in building feasibility, improving data infrastructure, developing skills, and creating organizational readiness so these become Quick Wins within twelve to eighteen months.

The same insurance company identified "predictive modeling for fraud detection across all claim types" as a Strategic Investment. High value because fraud costs them millions annually. Low feasibility because their data was fragmented across legacy systems, they lacked data science capability, and the models required sophisticated pattern recognition across complex scenarios. They knew this had to be a future capability, so they began the foundational work: consolidating data, hiring expertise, and running small pilot studies to learn.

Low Value, High Feasibility: These are "Efficiency Plays," things you could do easily but that don't move critical metrics much. Be selective here. Some efficient plays are worth doing because they build organizational comfort with AI, create demonstration cases, or solve annoying problems that consume disproportionate attention. Do not mistake activity for progress. Implementing five Low Value initiatives does not equal one High Value initiative.

"Automate meeting scheduling for executive team" was in this quadrant for one company. Easy to do, AI tools existed off the shelf, but the business value was minimal, maybe saving a few hours weekly for administrative staff. They implemented it because it was essentially free and gave non-technical executives direct experience with AI capability, making them more comfortable with larger initiatives. They did not celebrate it as strategic progress.

Low Value, Low Feasibility: These are "No-Go Zones" initiatives that are both difficult and don't deliver much value. Do not pursue these, no matter how technically interesting they might be. If vendors are pushing you toward something in this quadrant, it tells you whose interests they are serving.

For the insurance company, "natural language processing for analyzing handwritten claim forms from the 1990s" landed here. Low value because those old claims were largely resolved and represented tiny percentage of current workload. Low feasibility because handwriting recognition on aged documents is technically difficult and their archive systems made data access challenging. No business case justified the effort.

BUILDING YOUR AI INVESTMENT THESIS

Eventually someone will ask you to justify the investment. "Why are we spending so many millions on this? What is the return? How confident are you?" You need answers that are both honest and compelling, structured in ways that financial decision-makers can evaluate.

An investment thesis for AI should include five components:

The Problem Statement and Current Cost: This is where you explain what is broken and what it costs. Use the work from earlier in this chapter. "Unplanned equipment downtime costs us $5.4 million annually in lost production, emergency repairs, and expedited shipping to meet customer commitments. The frequency of failures has increased thirty percent over three years as equipment ages, and our current preventive maintenance approach is proving inadequate."

The Proposed Solution and Expected Outcomes: Explain what you are proposing and what you expect to achieve. Be specific but do not overpromise. "Implement predictive maintenance AI that analyzes sensor data to predict equipment failures before they occur. Expected outcomes: reduce unplanned downtime by sixty percent, shift from reactive to planned maintenance, extend equipment life by catching problems earlier. Conservative estimate: $3.2 million in annual savings from downtime reduction alone."

Investment Required and Timeline: Break down what you need to spend and when you expect returns. Include implementation costs, ongoing operational costs, and timeline to value. "Total investment: $800,000 for initial implementation over six months, including data infrastructure upgrades, AI platform, integration work, and training. Ongoing costs: $120,000 annually for platform maintenance and continuous model improvement. Expected timeline: value realization beginning at month nine, reaching full targeted savings by month eighteen."

Risk Assessment and Mitigation: Acknowledge what could go wrong and how you will address it. Financial decision-makers respect honest risk assessment more than overconfident predictions. "Key risks: sensor data quality may be insufficient for accurate prediction, requiring additional instrumentation ($150,000 contingency). Integration with existing maintenance systems may be more complex than estimated. Adoption by maintenance teams may be slower than planned. Mitigation: Six-week data quality assessment before full implementation. Phased integration approach with fallback to current systems. Intensive training and change management with maintenance leadership."

Success Criteria and Kill Gates: Define how you will know if it is working and when you would stop if it were. "Success criteria: sixty percent reduction in unplanned downtime, 75% accuracy in failure prediction with lead time of at least 48 hours, ninety percent adoption by maintenance staff within six months.

Kill gates: If after six months we have not achieved at least thirty percent downtime reduction, or if prediction accuracy is below fifty percent, or if maintenance staff adoption is below forty percent, we evaluate whether to continue, pivot, or terminate."

That structure gives decision-makers what they need: clear business logic, realistic expectations, honest risk assessment, and defined success criteria. It's investment-grade thinking, not hopeful speculation.

THE STRATEGY DOCUMENT: MAKING IT REAL

Strategy sitting in people's heads or scattered across meeting notes is not strategy, it is intention. Real strategy gets documented in ways that create accountability and enable execution.

Your AI strategy document does not need to be a hundred pages. In fact, it should not be. What it needs is clarity on critical questions and commitment from stakeholders. Ten to fifteen pages capturing what matters.

Section 1: Strategic Context (2 pages)

Connect AI initiative to broader business strategy. What are you trying to achieve as an organization? Where does AI fit? What business problems or opportunities do they address? This grounds everything that follows in business logic rather than technological enthusiasm.

Section 2: Current State Assessment (2 pages)

Where are you today in terms of AI maturity, data readiness, technical capability, and organizational readiness? Be honest. This isn't about impressing anyone; it's about establishing baseline reality so you can plan realistically.

Section 3: AI Vision and Objectives (2 pages)

Your North Star articulated clearly. What does success look like? For what are you optimizing? What constraints must you operate within? What's your timeline for achieving objectives?

Section 4: Priority Initiatives (3 pages)

Detail your top three to five initiatives based on your Value-Feasibility Matrix analysis. For each: business rationale, expected outcomes, resource requirements, timeline, success metrics, and responsible ownership. Three to five initiatives, not twenty. If you can't focus on a brief list, you can't execute effectively.

Section 5: Resource Allocation (2 pages)

Budget, people, technology investments, and timeline. Break down first-year costs versus ongoing costs. Identify where resources come from. If you're asking for net new budget, justify it. If you are reallocating from other initiatives, acknowledge what you're deprioritizing.

Section 6: Governance and Accountability (2 pages)

Who owns what? How will decisions get made? What's the cadence of review and adjustment? Who has authority to kill initiatives that aren't working? What metrics will you track and report to leadership? Governance sounds bureaucratic, but it is actually about creating clarity that enables speed.

Section 7: Risk Mitigation (1 page)

Top risks and your mitigation strategies. Do not bury risk, elevate it. Organizations that succeed with AI are often those that see risks most clearly and plan around them.

This document becomes your North Star, the reference point when priorities get fuzzy or when shiny new possibilities threaten to distract you from commitments you have made. Review it quarterly. Update it as you learn. Do not change it casually, because constantly shifting strategy is indistinguishable from having no strategy at all.

THE FOUNDATION YOU BUILD ON

Everything that comes after this chapter builds on the foundation you create here. Get the strategy right, ground it in business reality, connect it to problems worth solving, and you dramatically improve your odds of success. Skip this work, or rush through it, or let technology enthusiasm overwhelm business logic, and you are building on sand.

The manufacturing executive whose story opened this chapter eventually figured this out. Nine months after that awkward meeting where the CFO asked what problem they were solving, the company started over. New strategy process, grounded in business fundamentals. They identified their highest-value problem: quality defects caught late in production that cost millions in rework and delayed shipments. They built an investment thesis. They defined success criteria. They started with a careful pilot that proved value before scaling.

Two years later, they reduced quality defects by forty-two percent and saved $8 million annually. The technology they ultimately used was straightforward, nothing revolutionary. The strategic foundation made the difference. They knew what problems they were solving, why it mattered, and how they would know if they succeeded.

That is what this chapter equips you to build. Not a document that reads impressively but means little. A strategic foundation solid enough to support the implementation work ahead, clear enough to guide complex decisions, and honest enough to keep you grounded when hype and pressure push you toward unwise choices.

The next chapter tackles implementation. Before you move forward, make sure your strategic foundation can bear the weight. Once you start building, discovering your foundation is flawed becomes exponentially more difficult and more expensive.

CHAPTER 7

THE IMPLEMENTATION ROADMAP

The project plan looked beautiful. Twelve months, carefully sequenced phases, clear deliverables at each milestone. Data preparation in months one through three. Model development in months four through six. Testing and refinement in months seven through nine. Deployment in months ten through twelve. Neat, logical, comprehensive.

Eighteen months later, they were still in data preparation.

This is the dirty secret of AI implementation: the roadmap you start with bears almost no resemblance to the journey you actually take. Not because you planned badly, though maybe you did, but because implementation reveals problems you could not have known existed until you started solving them. Data that looked clean turns out to be inconsistent. Stakeholders who seemed aligned suddenly have fundamental disagreements. Technical approaches that evaluated perfectly in pilot hit inexplicable problems at scale.

The organizations that succeed don't have better initial roadmaps. They have roadmaps designed to accommodate learning and adaptation. They build in feedback loops, decision gates, and explicit permission to change direction when reality demands it.

Think of your roadmap less like turn-by-turn GPS directions and more like a hiking trail map. It shows the general route, the major landmarks, the difficult sections. You still need to watch where you are stepping, adjust to weather conditions, maybe take a detour around an unexpected obstacle. The map guides you without constraining you.

THE PHILOSOPHY OF ITERATIVE IMPLEMENTATION

Before we get into the phases and milestones, let us talk about the underlying approach because this determines whether your roadmap helps or hinders your success.

Traditional IT project management optimizes predictability. Define requirements completely upfront. Lock down the scope and execute according to plan. You then measure progress against the original timeline. This works fine when you know exactly what you are building and how to build it.

AI implementation requires a different philosophy: optimize for learning speed.

You cannot define all requirements upfront because you do not yet know what's possible, what's difficult, or what users will actually find valuable. You cannot lock down scope because you will discover new opportunities and new constraints as you go. You cannot execute rigidly according to plan because the plan will be wrong within the first month, not because you planned badly, but because you have learned things that change what the right approach looks like.

This does not mean chaos. It means disciplined iteration.

You set direction, not destination, and work in short iterations (cycles) with clear learning objectives. You build feedback mechanisms that tell you quickly when something is not working and celebrate changing direction based on what you learn rather than treating it as failure.

> A logistics company spent three months building an AI system to optimize delivery routes. The model was technically impressive, sophis-

> ticated algorithms considering traffic patterns, delivery time windows, driver availability, vehicle capacity. Then they tested it with actual drivers. It turned out that drivers had developed deep local knowledge about things the model could not capture. Details such as which customers always had packages to return, which loading docks were fastest at which times, which roads were nominally shorter but consistently slower due to construction or school zones.

The model's recommended routes were theoretically optimal but practically worse than what experienced drivers were already doing.

A waterfall approach would have treated this as failure and saved three months of wasted input. The company's iterative approach treated it as learning and they pivoted. Instead of replacing driver expertise with AI, they built a system that suggested improvements to driver routes while allowing drivers to override and explain their reasoning. The explanation data became training data that made the model smarter. Six months later, they had a hybrid system that combined AI optimization with driver knowledge, improving delivery efficiency by eighteen percent while increasing driver satisfaction.

This is what learning speed looks like. Small cycles, fast feedback, and willingness to pivot.

THE 90-DAY QUICK WIN STRATEGY

You need momentum. Not someday when the project is complete, you need it in the first ninety days, or organizational patience starts eroding.

Quick wins are not about implementing something trivial to claim victory. They are about demonstrating value fast enough to earn the time and trust required for longer-term transformation.

The successful twenty percent identify a small but meaningful problem they can solve in ninety days. Small enough to be completable quickly and meaningful enough that solving it matters to someone who influences budgets or strategic decisions.

What makes a good ninety-day target?

Scope: A sole use case for a single business process, not a platform that enables future use cases. You are solving a specific problem, not building infrastructure.

Data: You already have most of the data you need, or you can collect it within two weeks. You are not waiting for data warehouse modernization projects or enterprise-wide data governance initiatives.

Stakeholders: You have one clear business owner who will measure success and can make decisions without requiring approval from seventeen other people.

Technology: You can use existing tools and platforms, or you can deploy recent technology without enterprise architecture reviews, security audits, and procurement processes that take months.

Success metrics: You can measure improvement within thirty days of deployment, using metrics someone already cares about.

An insurance company targeted claims processing time for a specific type of claim: auto glass replacement. It was a narrow scope. They had three years of historical data already. The claims department director owned the problem and could measure success. They used a cloud-based ML platform that did not require new infrastructure. They could measure processing time immediately.

Ninety days later, they had reduced average processing time from four days to eighteen hours. Not transformational for the entire company, but transformational for that specific business problem. More importantly, they had demonstrated that AI could work, that the team could deliver, that the promised benefits were achievable.

This brought them credibility for bigger initiatives.

The quick win strategy is not about low-hanging fruit for its own sake. It is about proving the capability to execute before you ask for the budget and patience required for longer-term transformation. It is about turning skeptics into cautious believers and cautious believers into champions.

Pick something you can win fast, win visibly, and win in a way that matters to someone with influence.

YEAR ONE: BUILDING THE FOUNDATION

The first year feels slow if you are impatient. You want to be implementing AI everywhere, transforming business processes, demonstrating dramatic ROI. Resist this temptation. Year one is about establishing the foundation that makes years two and three possible.

Three parallel tracks run through year one: data, capability, and culture. You are working on all three simultaneously because they reinforce each other.

THE DATA TRACK: MONTHS 1-12

You need to know what data you have, where it lives, what condition it is in, and what work it needs done to it for any Ai initiatives you have. This sounds boring. It is boring but it is also the difference between success and failure.

Months 1-3: Data discovery and assessment

Inventory your data assets. Not just the obvious enterprise systems, but the departmental databases, the Excel files everyone uses, the third-party data sources and the data trapped in documents and emails.

A retail company discovered they had customer data in fourteen different systems. Their e-commerce platform, their point-of-sale system, their loyalty program database, their customer service system, their email marketing tool, their mobile app backend. Each system had slightly different customer information, none of them synchronized reliably.

You can't build effective AI on fragmented, inconsistent data. Before you can build any models, you need to understand what your data requirements are.

Assess data quality. Missing values, inconsistencies, errors, and timeliness issues. Be brutally honest. Data that's "good enough" for traditional analytics often isn't good enough for machine learning. Models trained on bad data make bad predictions, and unlike humans, they make those bad predictions at scale and speed.

Document data lineage.

Where does each piece of data come from?

Who owns it?

What business processes create or update it?

When was it last modified?

This becomes critical later when you need to debug model problems or ensure compliance with data regulations.

Months 4-6: Data infrastructure minimum

You don't need perfect data architecture. You need a minimum viable data infrastructure that supports your priority use cases.

This might mean creating a data lake for one business domain, not the entire enterprise. It might mean building API connections between three specific systems, not integrating everything with everything. It might mean establishing data quality rules for the datasets that matter for your initial AI projects, not implementing enterprise-wide data governance.

Perfection is the enemy. Build what you need for your first real AI implementations, knowing you'll expand and improve it later.

Months 7-12: Data pipeline automation

Manually preparing data for each AI model doesn't scale. You need repeatable, automated processes for extracting, transforming, and loading data that your AI systems will use.

This is where you build data pipelines, automated workflows that take raw data from source systems, clean it, transform it into formats your models need, and deliver it reliably. Initially you might have three or four pipelines supporting your priority use cases. That is fine. You are establishing patterns and practices you will replicate as you scale.

By the end of year one, you should have clean, accessible data for your priority business problems, infrastructure to support several AI use cases, and automated processes that can scale as you expand.

THE CAPABILITY TRACK: MONTHS 1-12

You need people who understand AI, can build AI systems, and can integrate those systems into business operations. You don't have enough of these people right now.

Months 1-3: Skills assessment and gap analysis

Who on your team already has relevant skills? Maybe your data analysts have statistical backgrounds. Maybe your software developers have worked with APIs and cloud platforms. Maybe your business analysts understand the business processes you're trying to improve.

Identify your gaps. You probably need people who understand machine learning algorithms. People who can engineer features from raw data. People who can deploy and monitor models in production and those who can translate between business requirements and technical solutions.

Months 4-6: Core team assembly

Build or acquire your core AI team. This is probably a mix of hiring, contracting, and upskilling existing people.

Hire for the skills you cannot build quickly. A senior ML engineer or data scientist with production experience can accelerate your timeline dramatically. They have made the mistakes you're about to make, know how to avoid them, and can mentor others.

Contract for specialized skills you need short-term but not long-term. Maybe you need computer vision expertise for one project but will not need it ongoing. Contract it rather than hiring permanently.

Upskill your existing people for the roles that benefit from deep business knowledge. Your best business analysts can become excellent AI product owners with training. Your strong software developers can learn Machine Learning Ops practices. Your data analysts can develop feature engineering skills.

Months 7-12: Capability building at scale.

Now you're expanding beyond the core team. You need business units to develop "AI literacy", not everyone needs to build models, but everyone needs to understand what AI can do, what it cannot do, how to work with AI systems, and how to identify opportunities.

Run workshops for business leaders on AI possibilities and limitations. Create lunch-and-learn sessions where teams share what they are learning. Develop internal case studies from your quick wins. Build an Ai Community of Practice where people working on AI initiatives can collaborate and share challenges and successes.

By year one's end, you should have a capable core team delivering projects, a broader organization that understands AI well enough to identify opportunities and work effectively with AI systems, and a learning culture that is continuously building capability.

THE CULTURE TRACK: MONTHS 1-12

Culture work starts immediately because cultural resistance kills more AI initiatives than technical challenges.

Months 1-3: Transparency and involvement

Communicate what you're doing and why. Not just top-down announcements, but conversations. What problems are you trying to solve? How might Ai affect distinct roles? What concerns do people have?

A manufacturing company made a deliberate choice early. They told production workers about plans to implement predictive maintenance before they told Wall Street. This was not legally required; it was culturally strategic. Workers heard directly from leadership that the goal was to reduce unexpected breakdowns, not reducing headcount. They heard it before rumors started. They heard it with specific details about how their roles might change, not vague assurances that everything would be fine.

This transparency brought trust that became valuable when implementation got difficult.

Months 4-6: Early involvement and feedback

Bring end users into the design process early. Not in superficial "we're gathering requirements" sessions, but in meaningful "help us understand your work" partnerships.

When a manufacturing company built their predictive maintenance models, they embedded a data scientist with the maintenance team for three weeks. She shadowed them, asked questions, learned their workflow. She discovered they already had intuitive prediction systems based on sounds, vibrations, and subtle performance changes that sensors were not capturing.

This ethnographic approach identified gaps between what the data showed and what experienced workers knew. The resulting models combined sensor data with worker input, creating a hybrid system that was more accurate and more trusted than either approach alone.

Months 7-12: Change champions and narrative.

Identify and empower your change champions, people in business units who are excited about AI possibilities, respected by their peers, and willing to advocate for innovative approaches.

These champions become your distributed change agents. They are the ones explaining to their colleagues why new processes exist, troubleshooting adoption problems, feeding back what is working and what is not.

You are also establishing the narrative. Not "AI is transforming everything," which creates anxiety. Something more specific to your context: "We're using AI to eliminate the most tedious parts of jobs so people can focus on work that requires judgment and expertise." Or "We're using AI to help us make better decisions faster, not to replace human decision-making."

The narrative matters because people interpret change through stories. Give them a story that is honest, specific to their context, and shows them a role in the future state.

YEAR TWO: SCALING WHAT WORKS

You've spent year one proving you can deliver, building foundation and generating momentum. Year two is about scaling success without scaling dysfunction.

The temptation is to launch ten new AI initiatives because now you know how to do this. Resist. You'll dilute focus, overwhelm your capacity, and probably fail at most of them. It is better to scale three things successfully than launch ten things that all struggles.

Months 13-18: Scale proven use cases

Take your year one successes and expand them. If you optimized routing for one logistics region, expand to three more regions. If you automate claims for one product type, expand to additional product types. If you implemented predictive maintenance for critical equipment, expand it to secondary equipment.

Scaling is not just deploying the same model more places. You are building organizational muscle for change. You are establishing patterns: how we prepare data, how we train teams, how we measure adoption, how we troubleshoot problems.

Each expansion teaches you something. The model that worked in region one needs adjustment for region two because business practices differ. Maybe the change management approach needs refinement because regional culture varies. This is valuable learning that makes your third and fourth implementations faster and smoother.

Months 19-24: Add complementary use cases.

Now you are expanding into adjacent problems where you can leverage the data infrastructure, technical capability, and organizational momentum you have built.

If your first implementations focused on operational efficiency, you now tackle customer experience problems. If you started with internal processes, maybe you add customer-facing AI features. You are diversifying value delivery while building on established foundations.

A financial services firm started year one with fraud detection, narrow scope, clear value, contained risk. Year two they expanded fraud detection to additional transaction types, but they also added a complementary use case: customer identity verification during account opening. Different problems, but leveraging the same data infrastructure, similar technical approaches, and organizational capability they had built for fraud work.

By the end of year two, you should have multiple AI capabilities in production delivering measurable business value, established patterns for how you implement AI successfully, and organizational confidence that this approach works at scale.

YEAR THREE: TRANSFORMATION ACCELERATION

Year three is where AI moves from "innovative projects" to "how we do business." You are not just implementing AI solutions; you are building an AI-enabled operating model.

Months 25-30: Platform and capability maturity

The project-by-project approach that got you through years one and two becomes limiting in year three. You need platform capabilities, reusable infrastructure, tools, and practices that make new AI implementations faster and cheaper.

This might mean building a model deployment pipeline that standardizes how you move from development to production. It might mean creating a feature store that lets multiple models share common data transformations.

It might mean establishing Machine Learning Ops practices that automate monitoring, retraining, and updates.

You are also maturing organizational capability. Your AI team members are now experienced practitioners. Your business units have people who have participated in multiple AI implementations. Your leadership team has learned to manage AI initiatives effectively.

Months 31-36: Business model innovation

Now you can start thinking more. Not just improving existing operations but creating new capabilities that were not possible before.

An insurance company spent years one and two improving internal operations, claim processing, underwriting efficiency, and fraud detection. By year three, they had the capability and confidence to launch usage-based insurance products that required real-time data processing and dynamic pricing models. This was not incremental improvement; it was a new business model enabled by the AI capability they had built.

You are not changing business models but fundamentally transforming how core operations work. A healthcare system moved from AI-assisted diagnosis in pilot programs to redesigned clinical workflows where AI support was standard practice for multiple specialties.

Year three is where your investment starts compounding. Each new implementation is faster because you have established patterns. Each success builds organizational capability that makes the next success more likely. You are not just implementing AI. You are becoming an organization that continuously improves using AI as a core capability.

MILESTONE CHECKLISTS

Roadmaps need concrete milestones, specific outcomes you will achieve by specific times. Not vague aspirations, but clear deliverables you can check off.

End of Month 3 Milestones

- Quick win project identified and scoped with business owner commitment.
- Data inventory completed for priority use cases.
- Core AI team assembled (even if partially contracted).
- Leadership alignment on year one priorities and success metrics.
- Initial communication completed with affected business units.

End of Month 6 Milestones

- Quick win project delivered and demonstrating measurable value.
- Data infrastructure supporting 2-3 priority use cases operational.
- First production AI model deployed and monitored.
- Business unit feedback mechanisms established.
- Lessons learned documented and shared.

End of Month 12 Milestones

- 3-5 AI use cases in production delivering measured business value.
- Data pipelines automated for priority use cases.
- Organizational AI literacy program launched and measuring engagement.
- Change champion network established in key business units.
- Year two roadmap defined with business owner commitments.

End of Month 18 Milestones

- Proven use cases scaled to multiple business units or geographies.
- Platform capabilities supporting rapid new model deployment.
- Established patterns for data preparation, model development, and deployment.
- Measured improvement in implementation speed from year one baseline.
- Year three transformation strategy defined.

End of Month 24 Milestones

- 10+ AI capabilities in production across multiple business domains.
- Self-service AI capability for business units with appropriate governance.
- Measurable organizational capability maturity improvement.
- Business case validated with actual ROI from delivered projects.
- Long-term AI roadmap integrated into corporate strategic planning.

END OF MONTH 36 MILESTONES

- AI embedded in core business processes as standard practice.
- New business capabilities or models enabled by AI operational.
- Continuous learning and improvement systems are mature.
- AI capability sustainable without heroic effort from specialists.
- Organization is viewed as AI-capable in market and talent competition.

THE KILL GATES

Not everything you start should finish. Successful organizations have discipline to stop projects that are not working rather than throwing good money after bad trying to rescue them.

Build explicit decision gates into your roadmap where you honestly assess whether to continue, pivot, or kill each initiative.

90-Day Gate: Is this worth continuing?

After ninety days, you should know whether the problem you are solving is as valuable as you thought, whether the data supports building a viable solution, and whether key stakeholders remain committed.

Kill signals: Business owner has lost interest. Data quality problems look unfixable without massive investment. Technical approach is not yielding acceptable performance. Organizational resistance is stronger than anticipated and leadership will not address it.

Six-Month Gate: Are we seeing the trajectory toward value?

You might not have measurable business impact yet, but you should see leading indicators. Are models performing adequately in testing? Are users engaging with early versions? Are you making steady progress or constantly firefighting unexpected problems?

Kill signals: Model performance is not approaching acceptable thresholds. Users reject early versions for fundamental rather than refinement reasons. Timeline has slipped repeatedly. You have pivoted approach three times. Budget overruns are substantial with no clear path to completion.

Twelve-Month Gate: Is actual value materializing?

By twelve months, you should have measurable business impact from at least some implementations. Not necessarily full ROI, but clear movement in target metrics.

Kill signals: Deployed models are not being adopted. Technical performance is acceptable, but business metrics are not moving. The value case has changed significantly enough that original justification no longer holds. Opportunity cost has become clear; this money would create more value elsewhere.

Having kill gates does not mean you are looking for reasons to quit. You are creating permission to quit when continuing does not make sense. This is liberating. Teams stop defending failing projects and start honestly assessing what is working. Resources get redirected from struggling initiatives to promising ones.

A pharmaceutical company killed three AI projects in year one despite those projects being technically successful. The models worked but business priorities had shifted, the value case weakened, and leadership chose to redirect resources to higher-impact opportunities. This was not failure; it was disciplined resource allocation.

MANAGING THE UNEXPECTED

Your roadmap will encounter problems you did not anticipate. This is normal. How you respond determines whether these problems derail you or just delay you.

When data problems surface: Do not wait for perfect data. Define minimum acceptable data quality for your use case and work toward that threshold. Often you can deliver value with imperfect data while parallel efforts improve data quality.

When technical approaches fail: Budget time and resources for technical experimentation. Not every approach works on the first attempt. Successful organizations try multiple technical approaches in parallel during discovery phases rather than committing fully to one approach too early.

When organizational resistance emerges: Do not fight battles you do not need to win immediately. If a business unit isn't ready for change, work with units that are. Success creates demand. Let early wins persuade late adopters rather than forcing adoption universally from the start.

When vendors underdeliver: Have contingency options. Do not depend entirely on a single vendor platform or partner until they've proven reliability. Build capabilities that are not locked into specific vendors so you can switch if necessary.

When leadership priorities shift: Maintain alignment between AI initiatives and current strategic priorities. Be willing to pause or kill AI projects when business strategy changes rather than continuing projects that no longer align with organizational direction.

THE ADAPTIVE ROADMAP

The best roadmaps are living documents. You review quarterly, assess progress honestly, update based on learning, and adjust priorities based on changing conditions.

Create a simple one-page roadmap that shows current quarter, next two quarters, and longer-term horizon. Each quarter, you move things forward, add new opportunities you have identified, remove things that no longer make sense, and adjust timing based on actual progress.

This quarterly review becomes the rhythm that keeps your AI initiatives aligned with business reality. It's where you celebrate successes, acknowledge failures, share learning, and recommit to priorities.

The roadmap serves three purposes: it guides execution for teams doing the work, it communicates progress and plans to leadership and stakeholders, and it creates accountability for delivering value rather than just completing activities.

Implementation without a roadmap is chaos. Implementation with a rigid roadmap that does not adapt to learning is expensive failure. Implementation with an adaptive roadmap that guides without constraining is how the successful twenty percent turn strategy into results.

You have built your strategic foundation. You have a roadmap that is ambitious but achievable, structured but flexible, detailed enough to execute but adaptable enough to accommodate learning.

Now the real work begins.

CHAPTER 8

THE CHANGE MANAGEMENT IMPERATIVE

The technical team had done everything right. Six months of meticulous work, data pipelines humming, models accurate to within acceptable thresholds, user interface polished and intuitive. The CFO signed off. IT gave the green light. The steering committee approved deployment.

Two weeks after launch, usage sat at eleven percent.

The system was not broken. Something else had failed, something the technical team hadn't even thought to build or evaluating or deploying. They had built a solution nobody wanted to use because nobody understood why they should, how it would help them, or what would happen if they did not.

This is where most AI transformations actually fail, not in technology, not in strategy, not in the budget. They fail in the quiet rebellion of people who simply choose not to change. They keep using the old spreadsheet, the old process, the old way of making decisions. Your brilliant AI system becomes expensive shelfware, and everyone involved wonders what went wrong.

Here's the uncomfortable truth: seventy percent of AI failures are change management failures masquerading as technical problems. We blame data quality, model accuracy, and the challenges of integration. These might be contributing factors, sometimes significant ones. They are rarely the reason your AI initiative delivers zero business value.

The reason is simpler and harder to fix people didn't change their behavior.

THE INVISIBLE FAILURE MODE

Change management sounds soft, which is why technical teams and analytically minded executives underinvest in it. They allocate ninety percent of budget to technology and ten percent to "training and communication," as if people are a minor implementation detail rather than the whole point.

Then they act surprised when adoption fails.

A healthcare system spent millions of dollars building an AI-powered clinical decision support system. The models analyzed patient data, suggested diagnoses, flagged potential drug interactions, recommended treatment protocols. Technically impressive. Clinical sound. Carefully validated against best practices and peer-reviewed literature.

Doctors ignored it.

Not because the recommendations were wrong, they were usually right. Not because the interface was clunky, the UX team had done decent work. Doctors ignored it because nobody had convinced them this system was solving a problem they actually had, nobody had involved them in designing how it would fit into clinical workflow, nobody had addressed their concerns about liability when AI recommendations contradicted their judgment, and nobody had explained what would happen to doctors who didn't use it versus those who did.

The organization had treated change management as a communications exercise, sent some emails, run some training sessions, and put up some posters. They had missed the actual work: understanding why people resist change, addressing the legitimate concerns underlying that resistance, redesigning workflows so new behaviors make sense, creating incentives that align with desired adoption, building trust through involvement rather than announcement.

It was technically perfect software that created zero clinical value because the human system rejected the technical system.

This pattern repeats across industries, geographies, and company sizes. The details differ; the fundamental failure mode stays constant. Organizations optimize for technical success while treating organizational change as an afterthought, then wonder why their successful technology fails to transform anything.

WHY PEOPLE RESIST AI

Before you can manage change, you need to understand what you are asking people to change and why they will resist.

Resistance is not irrationality or technophobia or generational backwardness. It is usually a rational response to legitimate concerns that leadership has not been adequately addressed.

Loss of autonomy and expertise

People spent years developing expertise in how they do their work. Now you are introducing a system that automates parts of that expertise or suggests they should follow AI recommendations rather than their own judgment.

This feels like devaluation. Their hard-won knowledge matters less. Their professional identity gets questioned. They are being told, implicitly or explicitly, that the algorithm knows better than they do.

A senior financial analyst had spent twelve years developing intuition about which deals were likely to close, which customers were credit risks, and which market signals mattered. The new AI system was fairly good at these assessments, sometimes better than human judgment. Using it wanted to admit his expertise didn't matter anymore.

He was not wrong to feel this way. His expertise did matter less in the new system. That is uncomfortable to acknowledge but important to address honestly rather than pretending nothing significant is changing.

Fear of job security

When you introduce AI that automates tasks people currently perform, they immediately wonder if you are automating away their jobs. Leadership says, "we're augmenting, not replacing," which might even be true. People hear "not yet."

They are not paranoid. They have watched automation eliminate jobs in other industries, other companies, and other departments. They have heard executives talk about efficiency gains and productivity improvements, euphemisms that sometimes mean fewer people doing more work.

A customer service organization deployed AI to manage routine inquiries, freeing agents to focus on complex customer issues. Leadership insisted this was not about headcount reduction. Then the next quarter's hiring freeze hit customer service harder than other departments because, after all, they would "improve efficiency with AI."

Agents noticed. Trust evaporated. Adoption of the next AI initiative cratered because people learned that cooperation with AI systems could make their own positions redundant.

Disruption to established workflows

People develop routines that work for them. Not necessarily optimal from a process efficiency perspective, but functional within the constraints they operate under. AI systems disrupt these routines, often in ways that want to make work harder rather than easier.

The new system might save ten minutes on one task but add fifteen minutes to another. It might reduce errors in one area while introducing new complications elsewhere. Net positive on paper, net negative in daily experience for the people actually doing the work.

Lack of transparency and trust

AI systems often work in ways people cannot see or understand. The algorithm makes a recommendation. Why? Based on what data? What would change its recommendation? Nobody knows, or nobody can explain it in terms that make intuitive sense.

When people do not understand how something works, they do not trust it. When they do not trust it, they do not use it, or they use it reluctantly while maintaining parallel manual processes "just to be safe."

An insurance underwriter needed to explain to applicants why their premiums were higher than expected. The AI pricing model said they were higher risk. Why? The model processed forty-seven variables through a neural network that even the data science team could not fully interpret. "The algorithm says so" is not an explanation that builds trust with customers or with underwriters who need to defend pricing decisions.

Change fatigue

Your AI initiative isn't the only change people are dealing with. There is the ERP upgrade, office relocation, the Digital Transformation, the new performance management system, and the updated compliance requirements. Each change individually might be justified. Collectively they are exhausting.

People have limited capacity to absorb change. When you introduce AI transformation on top of everything else, you are not just competing with the old way of working. You are competing with every other change initiative draining people's energy and patience.

A retail organization launched their AI inventory optimization system six months after implementing a new point-of-sale system, three months after reorganizing regional management structure, and two months before a planned merger. Store managers looked at another new system requiring new processes and made a rational calculation: focus on the changes that would get them fired if ignored, deprioritizing ones they could work around.

The AI system could be worked around. So, it was.

THE ADOPT FRAMEWORK

Managing change is not mystical. It is systematic work addressing predictable human responses to disruption. The ADOPT framework breaks this work into five sequential phases, each building on the previous one.

Align: Creating Shared Understanding

Before you announce anything, before you train anyone, you need alignment around why this change matters and what success looks like.

This is not top-down communication. It is dialogue. You are listening as much as talking, understanding concerns as much as explaining benefits, adjusting plans based on what you learn rather than just selling predetermined decisions.

Start with stakeholder mapping. Not just executive sponsors and project teams. Involve everyone whose work will change, everyone whose cooperation you need and everyone who could derail adoption if they choose to resist.

For each stakeholder group, understand their perspective. What do they care about in their work? What frustrates them currently? What might they fear about AI? What would make this change feel like improvement rather than imposition?

A manufacturing company mapped stakeholders for their predictive maintenance initiative: maintenance technicians, production supervisors, plant managers, reliability engineers, procurement teams, finance analysts. Each group had different concerns.

Technicians worried about job security and whether AI recommendations would override their expertise. Supervisors worried about production disruptions during implementation. Plant managers worried about capital budget implications. Engineers were excited about better data but concerned about workload during deployment. Procurement worried about vendor management complexity. Finance wanted clear ROI metrics.

The company did not ignore these concerns or paper them over with generic reassurances. They addressed each one specifically: technicians were involved in model development and given override authority; supervisors got phased rollout schedules that minimized disruption; plant managers got detailed cost-benefit analysis; engineers got additional contractor support during implementation; procurement got vendor scorecards; finance got quarterly ROI dashboards.

This took three months of meetings, workshops, and revisions to plans. It felt slow. It bought alignment that made execution dramatically faster than if they'd skipped this work and jumped straight to deployment.

Design: Building Change Into the Solution

Change management is not something you do after building the technical solution. It is something you design into the solution from the start.

This means involving end users in design, not just gathering requirements from them. Requirements-gathering is "tell us what you need and we'll build it." User involvement is "work with us to design how this should work."

The difference matters. When people help design the solution, they develop ownership. They understand why certain decisions were made. They have already started adapting their mental models to accommodate the new system because they helped shape what that system would be.

A financial services firm was building AI-powered fraud detection. They embedded a fraud analyst with the data science team for the entire development process. She brought real cases, explained what patterns mattered to human investigators, assessed models against edge cases she'd encountered, and helped design the alert interface investigators would actually use.

The resulting system was different from what the data science team would have built alone. It flagged fewer false positives because the analyst explained which transaction patterns looked suspicious algorithmically but rarely indicated actual fraud. It provided more context with each alert because she showed them what information investigators needed to make quick decisions. The interface matched investigator workflow because she helped design it.

When the system was deployed, fraud investigators trusted it immediately because one of their own had built it with data scientists. Adoption was ninety-three percent within the first month.

Design for gradual adoption, not big-bang deployment. People need time to build trust, develop new habits, and adapt to workflows. Forcing everyone to switch simultaneously creates crisis and resistance. Allowing gradual rollout lets early adopters become your champions and late adopters learn from peers rather than being forced by mandate.

Design feedback mechanisms into the system itself. People need ways to report problems, suggest improvements, and see that their feedback influences evolution. Without this, the system feels imposed and static. With it, the system feels collaborative and adaptive.

Operate: Enabling New Ways of Working

Now you are deploying, and operational realities hit. Training matters here, but not the training most organizations default to.

Typical AI training focuses on system features. "Here's how to log in. Here is the dashboard. Click this button to generate a report." This teaches people how to operate the technology. It does not teach them how to work differently in ways that create value.

Effective training focuses on the new workflow. "Here's what your Monday morning looks like now. Here is how you prepare for a customer call with AI support. Here's what you do when the AI recommendation contradicts your judgment."

Better yet, embed training in actual work rather than abstract classroom sessions. People learn by doing, with support available when they are hit by confusion or problems. This means having coaches available during early use, not IT support for technical problems, but expert users who can help people integrate the new system into their actual work patterns.

A sales organization did not run training sessions for their new AI lead-scoring system. They paired each sales rep with a "success coach" for the first month. Coaches were not tech support. They were experienced reps who had used the system during the pilot. They collaborated with reps, showing them in real sales situations how to interpret AI scores, how to adjust their outreach strategy based on scores, when to trust the AI and when to override it.

This was expensive, you need a lot of coaches for a large sales team. It worked. After one month, usage was eighty-seven percent and sales productivity had increased measurably because reps were not just using the system, they were using it effectively to change how they prioritized and approached prospects.

Make support easily accessible. People will hit problems, have questions, and need help. If getting support is difficult or slow, they will work around the system rather than learning to use it properly. If support is immediate and helpful, they will persist through the learning curve.

A global company established a Slack channel for their AI initiative where people could ask questions and get answers from the project team, from peer users, from anyone who knew the answer. Response time averaged under thirty minutes during business hours. People asked hundreds of questions, many basic, some sophisticated and got helpful answers quickly. This channel became a learning resource, a community builder, and an early warning system for problems needing attention.

Perform: Reinforcing Desired Behaviors

People adopt new behaviors when those behaviors are rewarded and old behaviors become harder or less rewarding.

Look at your incentive systems. If you are asking people to use AI but they are measured and rewarded based on metrics that do not account for the new way of working, you've created a conflict. People will optimize for what determines their bonus, their promotion prospects, and their manager's approval.

A pharmaceutical company deployed AI to accelerate drug discovery research. They asked scientists to use AI tools to generate hypotheses and identify promising compounds. Scientists were evaluated based on publication records and successful drug candidates. Neither metric rewarded using AI tools, and both metrics rewarded staying with established research methods scientists knew well.

Usage was terrible until the company changed evaluation criteria. They added metrics around research productivity; hypotheses evaluated per quarter and time from concept to candidate. These metrics made AI adoption strategically smart for scientists' careers. Usage jumped.

Create visible success stories. People need to see that using the new system leads to positive outcomes resulting in faster work, better results, recognition, and advancement. Early wins from early adopters should be celebrated visibly and connected explicitly to AI usage.

Do not just celebrate technical success. Celebrate the human behavior change that created business value. "The Northeast sales team increased close rates by fourteen percent by using AI led scoring to prioritize prospects" is better than "Our AI model achieved ninety-two percent accuracy." The first celebrates changed behavior and business results. The second celebrates technical achievement that might not connect to value in people's minds.

Address non-adoption directly. If people are not using the system, find out why. Not to punish them, but to understand what is not working. Sometimes they have legitimate reasons. The system does not work well for their specific situation, the workflow integration is clunky, or they lack clarity on expectations. Sometimes they are just resisting change. Either way, you need to know and respond appropriately.

Make old behavior harder. If you want people to use the new AI-powered process, make the old manual process require extra steps, approvals, or justification. Not to be punitive, but to create gentle pressure toward the new way while still allowing exceptions when necessary.

A procurement organization wanted buyers to use their AI vendor recommendation system. They did not mandate it, but they required buyers who selected vendors not recommended by the AI to document their reasoning. This took five minutes. Not onerous, but enough friction to make using AI recommendations the path of least resistance for routine purchases while preserving buyer judgment for complex situations.

Transform: Embedding Continuous Improvement

Change management does not end when adoption hits acceptable levels. The goal isn't just getting people to use AI; it is creating organizational capability to continuously improve how you use AI and adapt to new AI opportunities.

This means building learning into operations. Regular retrospectives where teams discuss what is working, what is not, and how to improve. Data on system usage and business impact that informs ongoing refinement. Mechanisms for people to suggest and implement improvements rather than waiting for centralized project teams to make changes.

A logistics company established monthly reviews for each AI system in production. Usage data, business impact metrics, user satisfaction scores, and open discussion about problems and opportunities. These were not status reports for management. They were working sessions where the people using the systems daily collaborated with technical teams to make things better.

Over twelve months, these reviews generated seventy-three improvements to their various AI systems, some technical, some workflow, and some training. Individually small, collectively substantial. The systems got better continuously because the people using them had agency to improve them.

Create a community of practice around AI. People working with AI across different business units can learn from each other, share solutions to common problems, and develop collective expertise that makes the organization smarter about AI over time.

This community should not be just technical specialists. Include business users, managers, executives, anyone whose work involves AI. The diversity of perspectives creates richer learning than homogeneous technical groups.

Celebrate learning, not just success. When someone discovers that an AI system does not work well for a particular use case, that's valuable learning. When a team experiments with a new way of using AI and it fails, that's valuable learning. Creating a culture where people share these learnings openly rather than hiding failures is how organizations get smarter faster.

BUILDING YOUR AI CHAMPIONS NETWORK

You cannot manage change for an entire organization by yourself or with a small project team. You need champions distributed throughout the organization who advocate for change, support peers, and make adoption feel achievable rather than overwhelming.

Champions are not necessarily your most senior people or your most technically sophisticated people. They are people who combine three characteristics: They're respected by peers, they are enthusiastic about the AI initiative, and they are willing to invest time helping others.

A regional bank identified thirty champions across their branch network for an AI-powered customer service initiative. These were not branch managers, managers were too busy and sometimes too distant from frontline work. They were experienced customer service reps, personal bankers, and loan officers. People who collaborated directly with customers and their peers asked for help in tough situations.

The bank gave champions first access to the AI tools, extensive training, and explicit time allocation to support peers. When the system rolled out branch by branch, champions were the first line of support. They could answer questions, demonstrate best practices, troubleshooting problems, and provide encouragement when people get frustrated.

This network created distributed change capacity. Instead of a small central team trying to support hundreds of employees, you had dozens of local champions providing help embedded in daily work.

Champions need support themselves. Regular meetings with the project team where they can share what they are hearing, surface problems, and get questions answered. They need recognition and often compensation for the extra work they are doing. They should be given clear authority to escalate issues that need central team attention.

Make the champion role desirable. People should want to be champions because it provides career development, visibility, expanded networks, and skill building. If being a champion feels like unpaid extra work with no benefit, you will struggle to find good people willing to do it.

The bank treated champion experience as professional development. Champions got first access to new features, advance notice of roadmap plans, and opportunities to influence product direction. They got visibility with senior leadership who attended champion network meetings. They got problem-solving and coaching skills that helped their careers beyond this specific initiative. Being selected as a champion was recognized internally as a positive career signal.

After two years, fourteen of the original thirty champions were promoted, several explicitly citing their champion experience as demonstrating leadership capability.

OVERCOMING THE "US VS. THE MACHINES" NARRATIVE

People fear AI will replace them. Sometimes this fear is justified, AI does eliminate some jobs. More often, the fear is overblown, but the narrative is so powerful it creates resistance even when job security is not threatened.

You need a counter-narrative that is honest and specific to your context.

The honest part matters. Do not promise "no job losses" if that's not true. If AI reduces headcount in certain roles, acknowledge this while explaining what will happen to affect people. Retraining? Transition support? Natural attrition rather than layoffs? People can manage difficult truths better than they can manage being misled and then blindsided.

The specific part matters because generic platitudes do not convince anyone. "AI augments human capability" sounds nice but doesn't tell people what it means for their actual job. "AI will handle routine data entry so you can spend more time on complex cases requiring judgment" is specific enough to evaluate.

A telecommunications company was implementing AI in their customer service operations. They were honest: routine technical support calls would increasingly be managed by AI, which would reduce the need for tier-one support staff. They were specific about what this meant: no layoffs, but attrition would not be backfilled in tier-one roles. Instead, they were expanding to two specialist roles and hiring for customer retention and upselling positions.

Currently tier-one staff could train for two specialist roles. Not everyone would succeed, specialist roles required deeper technical knowledge, but the company provided training and assessed people fairly. Those who did not want specialist roles could move to retention or sales positions, which paid better than tier-one support.

This was not painless. People had to change roles, learn new skills, and accept uncertainty. However, the company's honest and specific approach prevented the toxic "us versus machines" narrative from taking hold. People understood they were navigating a difficult transition, not fighting an existential threat to employment.

Frame AI is a tool that manages tedious work so humans can do more valuable work. This only works if it is true for your implementation. If it is not true, find a different framing that is honest.

Show do not just tell. Let people experience working with AI before you ask them to fully commit. When they see firsthand that AI oversees the parts of their job they do not enjoy while enhancing the parts that require human judgment, fear often transforms into enthusiasm.

An accounting firm had staff worried that AI would automate away audit work. They ran a pilot where auditors used AI tools to analyze financial statements, flagging anomalies and high-risk areas. Auditors discovered the AI eliminated hours of tedious preliminary analysis, letting them spend more time on the investigative work they found professionally satisfying such as understanding business operations, questioning unusual patterns, and applying judgment to ambiguous situations.

After the pilot, auditors became advocates for AI adoption because they had experienced it making their work better, not eliminating their value.

COMMUNICATION TEMPLATES THAT ACTUALLY WORK

Most AI transformation communications fail because they are either too vague or too technical. Leaders announce, "we're embarking on an AI transformation journey to become a data-driven organization." Nobody knows what this means or how it affects them. Or they explain the technical architecture and model capabilities. Nobody cares except other technical people.

Effective communication answers five questions for every audience:

What is changing and when?

Be specific. Not "we're implementing AI." Rather, "Starting in April, customer service reps will use an AI assistant that suggests responses to common customer questions. Rollout will happen team by team over three months, with training provided two weeks before your team goes live."

Why are we doing this?

Connect to business problems people recognize. "We're losing customers due to long wait times and inconsistent service quality. This AI assistant helps reps answer questions faster and more accurately, improving customer satisfaction while reducing rep stress from having to memorize hundreds of policy details."

How will this affect my daily work?

Describe actual workflow changes. "When a customer asks a question, you'll see suggested responses based on similar past questions. You can use these responses directly, modify them, or ignore them and write your own. The system learns from which suggestions you find helpful."

What is expected of me?

Be clear about requirements and choices. "You're expected to complete two hours of training, try using the system for at least two weeks, and provide feedback on what is working and what is not. You are not required to use AI suggestions for every response: use your judgment."

What support is available?

Make help accessible. "Your team champion is Sarah, who's been using this system for six weeks and can answer questions. Technical support is available via the #ai-support Slack/Teams channel, with response time under thirty minutes during business hours. Weekly office hours open on Thursdays at 2pm for anyone who wants firsthand help."

Create templates that different managers can customize for their teams rather than requiring everyone to write communications from scratch. Good templates make it easy for busy managers to communicate effectively.

For the customer service AI implementation, the company created a communication template that team leaders personalized with their team's specific rollout date, their designated champion, and examples relevant to their customer segment. This ensured consistent core messaging while allowing local customization.

TRAINING BLUEPRINTS FOR DIFFERENT LEARNING STYLES

People learn differently. Some want comprehensive manuals. Some want short videos. Some want firsthand practice. Some want to learn from peers. Effective training accommodates these differences rather than forcing everyone through identical programs.

Create a learning menu, not a mandatory curriculum. Core requirements everyone must complete, plus optional resources people can use based on their preferences and needs.

For an AI-powered inventory management system, the core requirement was a thirty-minute interactive tutorial covering basic functionality. Beyond that: detailed documentation for people who wanted complete reference materials, short video guides for specific tasks, weekly practice sessions for firsthand learning, peer shadowing opportunities for those who learned best by watching experienced users.

Different people used different resources. Everyone achieved competence, but through paths that matched their learning preferences.

Make training available when people need it, not just during scheduled sessions. Just-in-time learning is more effective than training that happens weeks before people use the system. People forget what they learned if they do not use it immediately.

Embed learning resources directly in the system when possible. Contextual help, tooltips, guided workflows. People can learn as they work rather than switching between the system and training materials.

A financial planning AI had "suggested next steps" built into every workflow. Unpracticed users could follow these suggestions to learn standard processes. Experienced users could ignore them and work their own way. This lets people learn at their own pace without feeling either abandoned or constrained.

Measure learning by competence, not by training hours completed. Can people perform the necessary tasks? Can they solve common problems? Can they explain to others how the system works? These matter more than whether they attended every training session.

THE FIRST 90 DAYS: CRITICAL PERIOD

The first ninety days after launch determine whether your change management succeeds or fails. This is when people form impressions, when problems surface, when resistance hardens or dissolves.

You need concentrated attention during this period. Not just monitoring technical performance, but actively supporting adoption, addressing problems quickly, celebrating early wins, and adjusting based on what you learn.

Expect problems and plan for them. Have capacity to respond immediately when issues arise. Nothing kills adoption faster than people encountering problems and getting no help or slow help. They conclude the organization is not serious about this change and revert to old behaviors.

A healthcare system assigned dedicated support staff for the first ninety days of their clinical AI deployment. Not regular IT support, people who understood clinical workflows and could troubleshoot both technical and process issues. When a physician encountered a problem at 10am, someone was there helping by 10:15. This responsiveness-built trust that the organization was committed to making this work.

Gather feedback aggressively. You need to know what is working and what is not, which requires asking constantly and making it easy for people to tell you. Anonymous surveys capture some things. Direct conversations capture others. Usage data shows what people do versus what they say.

The healthcare system sent weekly pulse surveys with three questions: What worked well this week? What caused problems? What would make this system more useful? Response rate was seventy percent because surveys took ninety seconds and people saw their feedback driving visible changes.

Make rapid adjustments based on feedback. Not everything. You cannot accommodate every preference or complaint. But when you see patterns, respond quickly. This shows people how their feedback matters and that the system is adaptive rather than static.

Week three, the pulse survey showed physicians frustrated by irrelevant alerts during certain procedures. Week four, the technical team adjusted alert logic to account for procedure context. Week five, physicians noticed the improvement. This responsiveness transformed physicians from skeptical users into collaborative partners who kept providing feedback because they saw it leading to improvements.

Celebrate early adopters and early wins. Make success visible. Share stories of people using AI effectively and achieving better outcomes. This creates social proof that using the system is normal and beneficial.

The healthcare system featured different physician champions in weekly emails, explaining how they used clinical AI and what value they got from it. Other physicians saw peers succeeding and wanted related results. Adoption accelerated through social contagion rather than just mandate.

WHEN CHANGE MANAGEMENT FAILS

Sometimes it fails despite good efforts. Usage stays low. Resistance hardens. Business value does not materialize because behavior does not change.

You need to make decision criteria for when to persist and when to pivot or stop.

Persist if: Problems are solvable with reasonable effort. Resistance comes from fixable issues, unclear communication, inadequate training, workflow friction, insufficient support. Usage is low but trending upward. Early adopters are successful and can become champions for broader adoption.

Pivot if: The system does not fit how people work, and changing the system is feasible. Resistance comes from legitimate concerns that can be addressed through redesign. The business problem is real, but your solution approach needs adjustment.

Stop if: The business problem is not actually valuable enough to justify the change required. Resistance reveals that your value proposition does not hold in practice. Usage stays near zero despite addressing known barriers. Early adopters are not successful or are not more successful than people using old methods.

A consumer goods company-built AI for sales forecasting. After four months, usage was seventeen percent and forecast accuracy had not improved. Investigation revealed the problem: sales managers did not trust AI forecasts more than their own judgment, and using the system added time without adding value. The company had built a solution to a problem that did not exist the way they thought it did.

They stopped the rollout, redesigned the approach based on deeper understanding of how forecasting actually worked, and relaunched six months later with a tool that supported manager judgment rather than replacing it. The second version succeeded because they had learned from the first version's change management failure.

MAKING CHANGE MANAGEMENT REAL

Here is what separates organizations that successfully manage change from those that just talk about it:

Successful organizations invest as much in change management as in technology. Not just budget but leadership attention, project timeline, success metrics. If your AI project allocates five percent of budget and timeline to change management, you are not serious about adoption.

Successful organizations involve users from the beginning. Not gathering requirements from them, but designing with them, testing with them, learning from them. By launch, users feel ownership because they shaped what is launching.

Successful organizations create distributed change capacity through champion networks, peer support, and embedded resources rather than relying on a small central team to manage change for hundreds or thousands of people.

Successful organizations measure behavior change and business impact, not just technical deployment. They track usage, proficiency, business metrics affected by the change. They celebrate changed behavior that drives value, not just technology that works.

Successful organizations adapt quickly based on feedback. They do not defend their original approach when evidence suggests it is not working. They treat early deployment as learning opportunity and make rapid adjustments.

Most fundamentally, successful organizations recognize that technology does not create value. Changed human behavior creates value. Technology enables behavior change, but only if you do the difficult work of understanding why people resist, addressing legitimate concerns, designing systems that fit how people work, supporting them through learning, and reinforcing desired behaviors.

Your AI initiative can be technically perfect and strategically sound. Without effective management changes, it will fail. With effective change management, even imperfect technology can drive substantial value because people are using it effectively and improving it continuously.

Technology is the easy part. People are the hard part. Organizations that succeed at AI understand this and act accordingly.

CHAPTER 9

BUILDING YOUR AI DREAM TEAM

The job posting asked for a unicorn. Machine learning PhD, ten years of production experience, expertise in natural language processing and computer vision, proficiency in five programming languages, demonstrated leadership in enterprise environments, and this was the kicker, willingness to work for a salary twenty percent below market rate because "we're building something special here."

Six months later, the position remained unfilled. The AI transformation initiative sat stalled because leadership believed they could not start without the perfect person. Meanwhile, three talented data analysts sat underutilized in the business intelligence team, a software engineer who'd taught himself Python and machine learning on weekends kept volunteering to help, and the company paid consulting fees to a vendor who was happy to keep the relationship going indefinitely.

This is one of the biggest self-inflicted wounds in AI transformation: the fantasy team problem. Organizations imagine they need a roster of rare specialists before they can do anything meaningful with AI. They write job descriptions for people who barely exist. They pass over capable people already in the organization because those people do not match the imaginary ideal. They convince themselves that AI requires an entirely new breed of professionals that they must acquire from outside.

The successful twenty percent figured something out: you do not need a dream team to start. You need a capable team that can learn. The dream team emerges through doing the work, not before it.

THE ROLES YOU ACTUALLY NEED

Let's start with reality. Different organizations at distinct stages of AI maturity need different capabilities. A company doing its first AI pilot needs different skills than one scaling AI across multiple business units. Pretending otherwise leads to either over-hiring for where you are now or under-staffing for where you are trying to go.

Think of team building in phases, matching capability to current needs while building toward future requirements.

Phase One: First Implementations

You are doing your first serious AI projects. You are not sure exactly what you're doing yet. You are learning as much as building. Here's what you need:

Someone who understands the business problem deeply. This person probably exists in your organization already. They are subject matter experts, maybe a senior analyst or experienced manager who knows the process you are trying to improve or the decision you are trying to make. They can articulate what success looks like in business terms and recognizes whether AI outputs make sense in context.

A retail company started their AI journey with their most experienced demand planner, someone who had spent eight years forecasting inventory across seasonal cycles, promotional events, and market disruptions. She couldn't build models, did not know Python, had never heard of gradient boosting. She knew what good forecasts looked like, what patterns mattered, when numbers that seemed right were actually suspicious. This knowledge proved more valuable than technical expertise in getting their first AI project to deliver actual business value.

Someone who can build and deploy models. You need technical capability, yes. This might be a data scientist, might be a machine learning engineer, might be a smart software developer who has learned ML. They do not need a PhD. They need to know how to work with data, build models that perform adequately, and deploy those models into production systems where real users can access them.

Early on, practical versatility beats specialized depth. You'd rather have someone who can do adequate data engineering, adequate modeling, and adequate deployment than three specialists who each do one thing brilliantly but cannot work together effectively.

Someone who is connected to business and technology. Call this role a product owner, an AI project manager, a business analyst. They translate between business needs and technical possibilities. They manage scope, timelines, and expectations. They keep projects focused on value delivery rather than technical perfection.

This person prevents projects from drifting into interesting technical problems that do not solve business problems. They are the ones asking uncomfortable questions: "That's impressive, but will it actually change the business metric we're trying to move? Can we deploy it in time to matter? Will people use it?"

That is, it for phase one. Three people, four. You don't need a center of excellence, a data science department, or an AI platform team yet. You need to do a few meaningful projects, learn from them, and prove value.

Phase Two: Scaling Proven Use Cases

You have been successful pilots. Now you are expanding with more use cases, more business units, more complexity. Your team needs to grow, but thoughtfully.

Specialized technical skills. You might now need people with deep expertise in specific areas: computer vision if you are working with images, natural language processing if you are analyzing text, time series forecasting if you're predicting future states. These specialists join as you encounter problems that require their expertise, not before.

Machine Learning Ops capability. As you move from pilots to production at scale, you need people who can build infrastructure for model deployment, monitoring, retraining, and management. This is different from building individual models. It is about creating systems that let you deploy and maintain many models efficiently.

An insurance company realized they needed Machine Learning Ops when they had seven models in production, each deployed and monitored differently. Every model requires custom effort to update. When they needed to retrain models because regulations changed, it took three months of manual work. They hired someone to build standardized deployment pipelines, automated monitoring, and systematic retraining processes. The next regulatory change was managed in two weeks because the infrastructure existed.

Prompt engineers. If you are working with large language models, you need people who understand how to craft effective prompts, chain prompts together for complex tasks, and integrate LLMs into workflows. This is a distinct skill. It is not traditional programming; it is a hybrid of technical understanding, language precision, and domain knowledge.

A legal services firm discovered this when they tried to use LLMs for contract analysis. Their software developers could integrate the APIs and build the interface. They couldn't get the LLM to extract the right information reliably because they didn't understand how to structure prompts, provide examples, or manage edge cases. They brought in someone with prompt engineering expertise who redesigned the entire interaction model. Accuracy jumped from seventy-one percent to ninety-four percent.

Data engineers. You are working with more data from more sources. You need people who can build reliable data pipelines, maintain data quality, and create infrastructure that supports multiple AI use cases simultaneously. This might have been adequate in phase one with one data scientist wearing multiple hats. In phase two, it becomes a specialized need.

Phase Three: Transformation at Scale

You are operating AI as a core capability across the organization. Now you need organizational infrastructure, not just project teams.

AI architects. People who can design systems that work together, establish standards and patterns, make technological choices that serve long-term needs rather than just immediate projects. They're thinking about how your AI capabilities fit together into a coherent platform rather than a collection of disconnected point solutions.

Change management specialists. You have learned by now that technology isn't your bottleneck; organizational adoption is. You need people skilled at managing change, building training programs, supporting users, and measuring adoption. These are not technical roles in the traditional sense, though they need enough technical understanding to explain AI systems to non-technical audiences.

Compliance, governance, and Ethic's specialists. At scale, you are making decisions that affect customers, employees, and business outcomes in significant ways. You need people thinking about fairness, bias, privacy, transparency, and risk. Not as afterthoughts, but as integral parts of how you build and deploy AI.

A financial services company promoted someone from their compliance team to focus full-time on AI governance.

She did not have a technical background, which turned out to be an advantage. She could ask the questions technical teams did not think to ask: "How do we explain this decision to a customer who's denied? What happens if the model treats different demographic groups differently? Who is accountable if the AI makes a costly mistake?" These questions prevented several problems that would have been expensive to fix after deployment.

BUILD VS. BUY VS. PARTNER: THE THREE-WAY DECISION

You need capabilities. You have three ways to get them: build by hiring and developing people internally, buy by acquiring companies or teams, or partners through contractors, consultants, or vendor relationships.

Most organizations get this decision wrong by defaulting on one approach for everything. They try to build every capability internally, or they outsource everything to vendors, or they hire contractors for all specialized needs. The right answer varies by capability, by organizational context, and by timeline.

Build when:

The capability is central to your competitive advantage. If AI in a particular domain is core to your strategy, you want that expertise in-house, under your control, continuously improving.

You need the capability long-term. Building takes time and investment. It only makes sense if you will need this capability for years, not months.

You have time to develop expertise. Building means hiring people with foundational skills and giving them time to develop specialized expertise in your context. If you need the capability immediately, building will not work.

The knowledge compounds internally. Some capabilities get better, the longer they are embedded in your organization because they accumulate context-specific knowledge. A data scientist who has worked in your business for three years knows things that even a more technically skilled contractor cannot replicate quickly.

Buy when:

You need capability quickly and in volume. Acquiring a team gets you up and running faster than hiring individuals, assuming you can find the right acquisition target and integrate them effectively.

The capability is proven and packageable. You are buying something that already works rather than building it from scratch.

The economy makes sense. This is rarely about AI core capabilities but sometimes about supporting functions or specialized niches.

Acquisitions are tricky. They succeed when you have clear integration plans and realistic expectations. They fail when you acquire people expecting them to continue performing in your organization the way they did in theirs, without accounting for culture clash, process friction, or key people leaving.

Partner when:

You need specialized expertise temporarily. You are working on a specific project that requires capabilities you will not need continuously. Computer vision expertise for a one-time implementation, for instance.

The capability is available and proven externally. Good partners exist, they have solved similar problems before and engaging them is faster than building.

You want to evaluate an approach before committing. Partnering lets you experiment with a capability to see if it creates value before making long-term investments in building it internally.

You need to move quickly. Partners can start immediately; hiring takes months.

Risk is in execution quality. Partners vary enormously in competence. You need rigorous evaluation and active management. The best partnerships involve knowledge transfer. Your team learns from the partner, so you build internal capability even while using external resources.

A manufacturing company partnered with a consulting firm for their first AI projects. The engagement was explicitly structured for knowledge transfer. Every external consultant collaborated with an internal counterpart. Decisions were made jointly with explanation. Documentation was comprehensive. After twelve months, the consultants rolled off and the internal team continued independently, having learned not just what to do but why and how.

THE COMBINED APPROACH

Smart organizations use all three approaches for diverse needs.

Core capabilities that differentiate you competitively: Build. Hire people, invest in their development, retain them long-term.

Specialized technical needs that recur but are not differentiating partners strategically with firms you can build ongoing relationships with, potentially bringing some of this capability in-house over time as volume justifies it.

Temporary specialized needs: Contract for specific engagements with clear deliverables and end dates.

Infrastructure and platform capabilities: Often this is where vendor solutions make sense, especially for well-established needs that aren't unique to your organization.

A healthcare system built their clinical AI team in-house; these were core capabilities requiring deep healthcare domain knowledge. They partnered with a specialized firm for natural language processing of medical records because they needed this capability but did not want to hire NLP specialists permanently. They used vendor platforms for model deployment and monitoring because this was infrastructure, they didn't need to build themselves. They contracted specific expertise for security audits and bias testing of high-stakes models.

This mix gave them the capabilities they needed without requiring them to hire every possible specialist or outsource everything.

THE CENTER OF EXCELLENCE THAT ACTUALLY WORKS

AI Centers of Excellence have become popular. Most of them fail, not because the concept is wrong but because organizations implement them wrong.

The failure pattern is predictable: create a centralized team of AI specialists, separate from business units, tasked with "driving AI transformation across the enterprise." The team builds impressive technical capabilities. They create beautiful demos. They struggle to deliver business value because they're disconnected from actual business operations, solving problems nobody really has or solving real problems in ways that don't fit operational reality.

Centers of Excellence that work are structured differently. They exist to enable business units to succeed with AI, not to do AI for business units.

THE HUB-AND-SPOKE MODEL

The central team is small, specialized, and focused on capability building. They establish standards, create reusable platforms and tools, provide expertise that individual business units can't justify maintaining, and help coordinate across initiatives so different teams aren't solving the same problems separately.

The real AI work happens in business units, with people who understand the specific context, problems, and constraints. They're building and deploying AI for their own operations, supported by the center but not dependent on the center to do everything.

A global logistics company structured their AI CoE with six people centrally: one focused on Machine Learning Ops infrastructure, one on data engineering standards, one on model governance, one on prompt engineering and LLM capabilities, one on vendor management, and one on change management and training. These six supported forty people distributed across regional operations, business development, and customer service who were implementing AI for their specific domains.

The distributed teams could move quickly because they owned their initiatives. The central team prevented duplication, established guardrails, provided specialized expertise, and facilitated sharing of learnings across units.

Clear value proposition

The center needs to provide value that business units couldn't achieve on their own. Otherwise, why would business units engage with the center rather than just doing their own thing?

Value might be: Technical infrastructure that would be inefficient for each business unit to build separately. Access to specialized skills that individual units can't justify hiring. Coordination that prevents teams from solving the same problems independently. Risk management and governance that protects the organization. Bargaining power with vendors through consolidated purchasing.

If the center can't articulate clear value, it becomes bureaucratic overhead that business units' route around.

Governance without gridlock

Centers of Excellence (CoE) often become centers of bureaucracy. Every AI initiative requires center approval. Processes multiply and speed dies.

Effective CoEs balance compliance and governance with autonomy. They establish guardrails, standards that must be met, risks that must be addressed, coordination points that require alignment, but within those guardrails, business units operate independently.

Think of it like building codes for construction. The code establishes safety standards, structural requirements, and approval checkpoints. Within those constraints, builders design and build freely. You do not need central approval for every design decision, just the ones that affect safety and code compliance.

An AI CoE might require: Security review for any model accessing customer data. Bias testing for models affecting hiring or lending decisions. Architecture review for new platform components. Knowledge sharing when multiple teams are working on similar problems. These are reasonable governance points. Requiring center approval for data preprocessing approaches, model selection, or feature engineering decisions creates gridlock without adding value.

Upskilling Your Existing Workforce

You don't need to hire anyone. Many of your best AI team members are already working for you; they just don't know if they could be doing AI work yet.

The question is not whether to be upskilled, you must, because AI capabilities are becoming as fundamental as Excel proficiency was twenty years ago. The question is how to upskill effectively rather than wasting money on training that does not stick.

THE 70-20-10 DEVELOPMENT MODEL

This framework, originally from leadership development, applies perfectly to AI skill building.

Seventy percent of learning happens through challenging work assignments. People learn AI by doing AI projects, making mistakes, solving real problems, getting feedback. Training courses provide foundation, but you cannot learn AI from courses alone any more than you could learn to ride a bike from a manual.

This means you need to create opportunities for people to work on real AI projects while they are still learning. Not after they are fully trained during the learning process. Small projects, low-stakes problems, supported by more experienced people, but real work that matters.

A retail bank identified ten analysts who showed interest in AI. Rather than sending them to six months of training before they could contribute, the bank paired each analyst with a data scientist and gave them joint responsibility for a real project. The analyst brought business knowledge; the data scientist brought technical expertise. They worked together, the analyst learning constantly through practical application while simultaneously contributing value through domain knowledge.

After six months, those analysts could do meaningful AI work independently. The ones who'd gone through traditional training-first approaches took longer to become productive and retained less of what they had learned.

Twenty percent of learning comes from social learning, working with others, getting feedback, observing how experienced people approach problems. This is why mentorship, pair programming, and communities of practice matter.

Create structures for experienced AI practitioners to collaborate with people developing skills. Not formal training relationships necessarily, just proximity and collaboration. Learning happens through osmosis as much as instruction.

Ten percent of learning comes from formal training, courses, certifications, workshops. This provides foundational knowledge and structured skill development. It is necessary; it is not sufficient.

The mistake is inverting these ratios. Organizations send people to weeks of formal training, then expect them to apply it independently without opportunities for learning through doing or learning from others. The formal knowledge fails to stick without practical application and social reinforcement.

Targeted skill development, not universal AI literacy

Dissimilar roles need different AI capabilities. Don't train everyone on everything; target development to what dissimilar roles actually need.

Executives need AI literacy, enough understanding to make informed decisions about AI investments, recognize good and bad AI opportunities, ask the right questions, and provide effective oversight. They do not need to understand transformer architectures or gradient descent algorithms.

Business analysts and domain experts need enough AI knowledge to identify opportunities, collaborate effectively with technical teams, and evaluate whether AI solutions are solving the right problems. They might need basic understanding of how different AI approaches work, what data is required, and what makes projects succeed or fail. They do not need to build models themselves.

Data analysts and business intelligence professionals might develop into AI practitioners with training in machine learning fundamentals, feature engineering, model evaluation. They already understand data and statistics; they are learning to apply those skills with AI tools.

Software engineers need to understand how to integrate AI into applications, deploy models, monitor performance, and oversee errors. They are learning Machine Learning Ops and production AI systems, not necessarily model development.

A pharmaceutical company mapped roles to needed AI capabilities and created targeted learning paths for each. Research scientists needed to understand how AI could accelerate drug discovery but not how to build deep learning models, they partnered with specialists for that. Clinical trial managers needed enough knowledge to identify where AI could improve trial design and patient monitoring. Regulatory affairs are needed to understand AI governance and explainability. Marketing analytics needed firsthand ML skills.

Different paths for diverse needs, rather than one-size-fits-all AI training.

LEARNING BY TEACHING

The best way to solidify learning is to teach others. As people develop AI capabilities, create opportunities for them to share knowledge, internal workshops, lunch & learn, documentation, mentoring others starting their learning journey.

This reinforces their own learning while distributing knowledge across the organization. It identifies who has really mastered material versus who has just gone through training motions. People who can explain something clearly have understood it deeply; people who cannot explain it have not really learned it yet regardless of what courses they have completed.

Vendor Evaluation and Management

You will collaborate with vendors. Technology platforms, consulting services, specialized capabilities you are not building internally. Managing these relationships well is the difference between vendors who accelerate your AI success and vendors who become expensive distractions.

The evaluation scorecard

Do not evaluate vendors on demos. Demos are performance art, carefully rehearsed to show capability in ideal conditions. Evaluate evidence of real-world performance with clients facing similar challenges in similar contexts.

Proven relevant experience: Have they solved similar problems for similar organizations? Not just "we've done AI projects" but specific evidence of success in your industry or with your case type.

Ask for references and call them. Ask specific questions: Did they deliver what they promised? Did it work in production or just in demos? How well did they manage problems? Would you hire them again?

Technical capability: Do they have the specific technical skills you need? Can they demonstrate depth in relevant areas? Are they using proven, production-ready approaches or experimenting with bleeding-edge techniques that might not be stable?

Be skeptical of vendors using the latest, hottest AI approaches for everything. Sometimes the newest techniques are genuinely better. Often, well-established methods work fine and are more dependable.

Cultural fit: Will they collaborate effectively with your team and organization? Some vendors are genuinely collaborative, treating your success as their success. Others are transactional, delivering contracted scope and nothing more. Others are empire-building, trying to make you dependent on them for everything.

Cultural fitness matters more than it seems. A technically excellent vendor that cannot work effectively with your team will create friction that undermines value.

Knowledge transfer approach: Are they trying to build your capability or build dependency? Good vendors want you to learn from them so you can eventually do more independently. Vendors optimizing for ongoing fees want you to stay dependent.

Ask explicitly: "What will our team learn through this engagement? How are you structuring work to transfer knowledge? When you leave, will we be able to maintain and extend what you have built?"

Pricing and contract structure: Are incentives aligned? Does the vendor make money by delivering value or by maximizing hours? Are there success-based components to pricing or only time-and-materials?

Aligned incentives do not guarantee success, but misaligned incentives almost guarantee problems. If a vendor profits more when projects take longer or require more ongoing support, they are not incentivized to build your independence.

Managing the relationship

Hiring a good vendor is the beginning, not the end. You need active management to get value.

Insist on working together, not being handed finished solutions. You want your team embedded in the work, learning the approaches, understanding the decisions. This slows things down slightly; it dramatically increases the value you retain when the vendor leaves.

> A media company engaged consultants for recommendation engine development. They could have let the consultants build everything and hand over completed code. Instead, they had internal developers pair-program with consultants throughout. The process took about twenty percent longer. When consultants leave,

> the internal team could maintain, troubleshoot, and enhance the system because they had built it together rather than just receiving it.

Demand documentation that enables future independence. Code comments, architecture documentation, decision logs explaining why certain approaches were chosen. This feels bureaucratic; it is essential for being able to maintain and extend work after vendors leave.

Have clear decision rights. What decisions can vendors make independently? What requires your approval? What requires joint decision-making? Ambiguity here creates either gridlock (everything requires ten meetings) or vendors making decisions you should be making.

Measure value, not activity. Do not track how many hours vendors work or how many meetings they attend. Track whether they are delivering business value, building your capability, and solving the problems you hired them to solve.

When to fire vendors

Sometimes vendor relationships are not working. Recognizing this quickly and acting decisively saves money and time.

Fire vendors when: They are not delivering promised capabilities or results despite adequate time and resources. They are creating dependency rather than building capability. They're more focused on selling additional services than delivering on current commitments. Cultural fitness problems are undermining collaboration. You have learned enough to do this work internally more effectively.

Do not fire vendors just because projects are hard or hitting obstacles. AI projects encounter problems; that is normal. Fire them when they are handling problems poorly, not learning from setbacks, or not being honest about challenges.

THE ROLES YOU DEFINITELY DON'T NEED

Part of building the right team is knowing that for which not to hire.

The Chief AI Officer who is not operational

Some organizations create executive AI roles for people who provide vision and strategy but do not roll up their sleeves in actual implementation. Vision without execution is expensive speech-making. If you are creating an executive AI role, that person should be deeply involved in implementation, not just strategy.

AI specialists for every technique

You don't need separate specialists in computer vision, NLP, reinforcement learning, generative AI, and every other AI subdomain unless you're actually working on problems requiring all of these. Hire for the problems you are solving, not for comprehensive coverage of AI techniques.

Full-time researchers

Unless you are a technology company where AI research itself is a product or competitive advantage, you do not need people doing research. You need people implementing proven approaches effectively.

Massive data science teams before you have real use cases

Organizations sometimes hire dozens of data scientists and then look for problems for them to solve. This is backward. Hire as you identify significant business problems worth solving, not in anticipation of someday finding uses for AI people.

BUILDING FOR WHAT YOU WILL BECOME

Here is the nuance: you are building a team not just for current needs but for where you're heading. This does not mean hiring for future capabilities you do not need yet. It means hiring people who can grow as your needs evolve.

Look for: Learning orientation over current expertise. Someone who is continuously learning new capabilities is more valuable long-term than someone with deep current expertise but no growth trajectory.

Versatility over narrow specialization early. In early phases, people who can contribute across multiple areas are more valuable than specialists, even if specialists are more technically deep.

Business sense alongside technical skills. People who understand both technology and business create more value than technically brilliant people who cannot connect their work to business outcomes.

Collaboration and teaching ability. As your AI capability matures, the people who can help others learn become force multipliers worth far more than their individual contribution.

The dream team is not who you hire on day one. It is who your team becomes through doing demanding work together, learning continuously, and building capabilities that compound overtime. Focus less on finding perfect people and more on building systems that help good people become great at AI through practice, support, and challenge.

That impossible job posting for a unicorn data scientist? A mid-sized technology company finally rewrote it. Instead of demanding someone who could do everything, they defined the actual problem they needed solved in the next six months, identified the minimum capabilities required, and added "demonstrated ability to learn quickly and work collaboratively."

They hired someone with solid fundamentals but not all the specific experience they had originally demanded. Paired them with business experts. Gave them a real problem to solve with support from the team. Twelve months later, that person was leading AI initiatives across three business units and mentoring others.

Not because they were a unicorn when hired. Because they were put in a position to grow into the role the organization needed, supported by a team structure that enabled learning through doing.

That is how you build your AI dream team, not by finding perfect people but by creating the conditions for capable people to become exceptional through meaningful work.

CHAPTER 10

THE DATA FOUNDATION

The presentation had sixty-three slides about their new AI strategy. Ambitious use cases, impressive vendor partnerships, detailed implementation timelines. The board approved the budget. Everyone left feeling energized about the transformation ahead.

Three months in, the data scientist they had hired asked a question nobody had thought to answer: "Where's the data?"

Turns out, customer transaction records lived in a system from the 90's that nobody knew how to export properly. Product information existed in seventeen different spreadsheets maintained by seventeen different people with seventeen different naming conventions. Sales pipeline data was entered manually, inconsistently, and often weeks after deals closed. The marketing team had their own database that did not talk to anyone else's systems. Finance had theirs. Operations had three.

This pattern repeats itself consistently. Organizations get excited about AI capabilities, make ambitious plans, assign budgets, hire talent. Then someone tries to actually build something and discovers their data situation makes it impossible.

Here is the part that stings: everybody involved was smart, experienced, capable. They had launched successful initiatives before. They just assumed data would be there when needed because data is always there, right? We are swimming in it. Big data, data lakes, data-driven decision making. How hard could it be?

Turns out data being abundant and data being usable for AI are entirely different conditions. You can have petabytes of data and still be unable to train a model because that data is dirty, scattered, inconsistent, poorly labeled, legally restricted, or just fundamentally not capturing what you need to predict.

The successful twenty percent learned this early. They did not start with AI strategy. They started with data reality.

WHY DATA STRATEGY DETERMINES AI SUCCESS

AI is fundamentally pattern recognition. Feed the algorithm examples, it learns patterns and applies those patterns to new situations. Simple concept. The entire thing lives or dies based on whether your data actually contains learnable patterns that matter for your business problem.

Think about what machine learning needs: large volumes of relevant data, consistent and clean enough that patterns emerge rather than getting obscured by noise, labeled or structured so the algorithm knows what it's looking for, accessible when and where you need it, legally and ethically usable for your intended purpose.

Now think about your data reality. It probably looks different.

A regional bank wanted to use AI for loan default prediction. It was a sensible use case. They had twenty years of loan data. Sounds promising. Closer inspection revealed problems: loan application data was complete and consistent, but only for the last five years because they had changed systems. Pre-2020 data existed but in formats that required manual interpretation. Default data was accurate but did not capture why loans defaulted. There were problematic macroeconomic conditions, borrower circumstances, and loan structure issues. Without understanding why, the model could not distinguish between several types of risk. Employment verification data was incomplete because verification requirements had changed over time. Income data was self-reported until 2018, verified after that.

They had data. They did not have data that would train a dependable default prediction model without substantial preparation work.

This is not unusual. It is typical. The organizations that succeed with AI accept this reality and do the unglamorous work of building data foundations before they try to build AI solutions. The ones that fail either do not understand the data requirements or understand them but convince themselves they can skip this work and figure it out later.

THE MINIMUM VIABLE DATA INFRASTRUCTURE

You do not need perfect data infrastructure to start with AI. You absolutely need minimum viable data infrastructure, and you need to be honest about whether you have it.

Minimum viable means: data you need for priority use cases exists and is accessible, quality is good enough that models can learn real patterns rather than artifacts of bad data, you have processes to maintain and improve data over time, and you understand what data you can legally and ethically use.

Notice what is not in that definition: enterprise data warehouse, data lake, master data management system, comprehensive data governance program. These might be worthwhile eventually. They are not prerequisites for starting AI work.

ACCESSIBLE DATA FOR PRIORITY USE CASES

You have identified high-value AI opportunities through your strategic planning. Now: can you actually access the data those use cases require?

Sometimes yes. A manufacturing company wanted predictive maintenance for critical equipment. They already had sensor data flowing from equipment into a monitoring system, maintenance logs in a structured database, failure incidents documented consistently. The data existed and was accessible. They could start building models.

Sometimes no. A professional services firm wanted to predict project profitability early in engagement. They had historical project data, but it was scattered across billing systems, project management tools, timesheets, expense reports, and email. Nobody had connected these data sources. Nobody had structured the information consistently. Extracting what they needed would require months of integration work before any modeling could begin.

The second situation is not a reason to abandon the use case. It is a reason to do data infrastructure work first, then AI work. Or to choose a different initial use case where data is already accessible while you build infrastructure for harder problems.

QUALITY SUFFICIENT FOR LEARNING

Data does not need to be perfect. It needs to be good enough that real patterns are stronger than noise.

What constitutes "good enough" depends on what you're predicting and how much data you have. More data compensates somewhat for lower quality because real patterns emerge across volume even if individual records are noisy. Higher quality data means you can work at less volume.

A telecommunications company had customer service interaction data going back eight years. Impressive volume. Quality was mixed. Early years had inconsistent categorization of issue types. Text fields had spelling errors, abbreviations, and incomplete information. Resolution codes did not always match what happened. Timestamps were sometimes off due to system issues.

They could still build useful models because patterns were strong enough to emerge despite noise. Calls about billing issues had recognizable patterns in customer language even when issue categories were wrong. Resolution time patterns showed up even when some timestamps were inaccurate. The models were not as good as they could be with perfect data, but they were good enough to add value.

Quality threshold varies by use case. High-stakes applications like medical diagnosis or financial fraud need higher quality than lower-stakes applications like content recommendations. You are making a judgment call about whether quality is adequate for your specific context.

MAINTENANCE AND IMPROVEMENT PROCESSES

Data does not stay good by itself. It degrades. Systems change, business processes evolve, people get sloppy about data entry, edge cases accumulate. You need processes that monitor data quality and fix problems before they break your AI systems.

An insurance company deployed models for claims processing. Initially accurate. Six months later, accuracy dropped significantly. Investigation revealed several issues: the claims system had been updated, changing how certain fields were populated. A major product line had been introduced with claim characteristics that did not fit existing categorization. Two regional offices have developed local data entry practices that deviated from standard process. Nobody had caught these changes because nobody was systematically monitoring data quality.

They implemented monitoring: automated checks for data distribution shifts, manual reviews of sample records, feedback loops from model performance back to data quality issues. When problems emerge, they could identify and fix them quickly rather than discovering those months later through degraded model performance.

LEGAL AND ETHICAL CLARITY

You can have abundant, accessible, high-quality data and still not be able to use it for AI if legal or ethical constraints prevent it.

Privacy regulations matter. GDPR in Europe, CCPA in California, and similar regulations globally restrict how you can use personal data. You need to understand what data you are allowed to use for what purposes, what consent requirements apply, and what documentation you need.

Ethical considerations matter even when legal requirements do not prohibit use. Just because you can use data does not mean you should. Using data in ways that feel invasive, discriminatory, or manipulative creates risks even if technically legal.

A retail company had detailed purchase data including what customers bought, when, where, and increasingly through loyalty programs, who they were. Legally they could use this data for AI-driven marketing. Ethically they needed to consider: Are we making inferences about sensitive attributes like health status or financial distress? Are we treating different customer segments differently in ways that could be discriminatory? Are we transparent about how we are using customer data? Could our use of this data damage customer trust if it became public?

They established guidelines: no use of purchase data to infer health conditions or financial difficulties. No differential pricing based on inferred demographic characteristics. Clear disclosure with customers about how their data enabled personalized experiences. Regular ethics reviews of new AI use cases.

These constraints limit some AI opportunities. They protected the company from reputational and regulatory risks that could have been far more costly than missed revenue.

DATA ARCHITECTURE WITHOUT COMPLEXITY

Data architecture sounds intimidating. It conjures images of complex diagrams, endless meetings, multi-year programs, consultants billing by the hour.

Forget all that. You need just enough data architecture to support AI work without creating complexity that slows everything down.

DEFINED AND APPOINTED DATA OWNERS

During an AI transformation, Data Owners become one of the most important roles in the enterprise because they ensure the organization's most valuable asset, its data, is accurate, governed, secure, and strategically aligned to business outcomes.

AI systems are only as good as the data they consume, and Data Owners provide the clarity, authority, and direction needed to define what "good" looks like. They establish data quality standards, enforce compliance, and decide how data can be used ethically and responsibly across the enterprise. By setting policies, prioritizing investments, and ensuring that data reflects the true needs of the business, Data Owners eliminate ambiguity and prevent AI initiatives from being driven by poor, biased, or inconsistent information.

Ultimately, they serve as the bridge between business strategy and technical execution, ensuring that AI delivers real, measurable value rather than operating as a disconnected or risky experiment.

START WITH USE CASE REQUIREMENTS, NOT COMPREHENSIVE DESIGN

Traditional data architecture starts with a target state design: how data should be organized, integrated, and governed across the entire enterprise. Then you spend years migrating toward that vision.

AI-focused data architecture starts with: what data do our priority AI use cases need? How can we make that data available quickly? What patterns can we establish now that we can replicate later?

> A logistics company needed package tracking data, route information, delivery times, and customer feedback for their first AI projects around route optimization and delivery prediction. Instead of designing a comprehensive data platform for all logistics operations, they built pipelines for these specific data sources. They got them working reliably and documented the patterns. When they added new AI use cases, they replicated these patterns for new data sources.

Over time, this evolved into more systematic data architecture. They had not avoided architecture; they had built it incrementally based on real use case needs rather than theoretical completeness.

THE THREE-LAYER PATTERN

Effective data architecture for AI typically has three layers, whether you're building something simple or sophisticated.

Source data layer: This is where data lives operationally. Your transaction systems, databases, applications, external data sources. You generally don't change these systems for

AI. They are doing their jobs supporting business operations. You extract data from them rather than modifying them.

Integration and preparation layer: This is where data from various sources gets combined, cleaned, and transformed into formats AI can use. Sometimes called a data lake or data warehouse, though those terms carry baggage about specific technologies. The key function is taking messy operational data and turning it into clean, integrated data sets ready for analysis and modeling.

Model- ready data layer: This is data specifically prepared for machine learning. Features engineered, training and test sets created, labeled data organized, formats optimized for whatever ML framework you are using. Some organizations skip this layer, preparing data directly from the integration layer for each model. This works initially but becomes inefficient as you scale.

CHOOSING TECHNOLOGY PRAGMATICALLY

You need to make technological choices: what platforms, what tools, what infrastructure. Make these decisions based on practical considerations, not vendor marketing or technology trends.

Consider: what does your team already know how to use? Learning new platforms has costs in time and mistakes. Building familiar technology accelerates delivery even if that technology isn't theoretically optimal.

What integrates well with systems you already have? Adding technology that plays nicely with your existing infrastructure is easier than technology that requires extensive custom integration.

What scales to your actual needs rather than hypothetical future needs? Over-investing in scalability you will never use wastes money. Under-investing in scalability, will definitely lead to expensive rebuilds later. Be realistic about what scale you'll reach.

What flexibility do you need for future unknowns? AI technologies evolve quickly. Platforms that lock you into specific approaches become liabilities. Some flexibility premiums are worth paying even if you are not certain you will need them.

A financial services firm evaluated data platforms for AI. Their data team knew SQL and Python well, had limited experience with newer big data tools. They chose a cloud data warehouse that supported SQL for data manipulation and Python for ML work. Not the most innovative choice, not the one recommended by analysts for "enterprise AI at scale." It was the choice that allowed their team to be productive immediately and handle their actual scale requirements without forcing them to learn entirely new technology stacks.

Two years later, this pragmatic choice had enabled them to deploy eleven AI applications. The team that chose the trendier, more complex platform was still building expertise and had deployed three applications.

DATA GOVERNANCE WITHOUT BUREAUCRACY

Data governance has a reputation for being bureaucratic overhead that slows everything down. This reputation is often deserved. Organizations create governance committees, approval processes, documentation requirements that turn simple data access into month-long exercises.

You need governance. You do not need bureaucracy. The difference is that good governance enables appropriate data use while preventing misuse. Bureaucracy just makes everything harder without making anything safer.

GUARDRAILS, NOT GATES

Think of governance as establishing guardrails that keep people safe while letting them move fast, rather than gates that stop people until someone gives permission.

Guardrails look like: automated data classification that tags sensitive data and restricts who can access it. Standard patterns for common data operations that people can follow without special approval. Clear criteria for what require review versus what is pre-approved. Self-service access to appropriately restricted data rather than ticket-based request processes.

A pharmaceutical company needed to balance data access for AI development with strict privacy and regulatory requirements. They classified all data into sensitivity tiers. Tier 1 data (de-identified research data) was accessible to anyone in the company for analysis. Tier 2 data (operational data with some sensitive elements) required basic training certification to access. Tier 3 data (extremely sensitive patient or commercial information) required specific project justification and security controls.

This classification was automated where possible, manual where necessary, but always clear. Researchers knew what they could access immediately and what required additional steps. They did not need to ask permission for Tier 1 data; they just used it. For Tier 2 and 3, the requirements were explicit and the approval process streamlined.

Most AI work used Tier 1 data, which had no access friction. The minority of projects needing sensitive data had clear paths to get it without bureaucratic delays.

DOCUMENTATION THAT SERVES REAL PURPOSES

Governance often generates documentation requirements: data dictionaries, lineage documentation, use case approvals, impact assessments. Some of this documentation is genuinely valuable. Most of it is theater.

Valuable documentation answers questions people have:
What does this data represent?
Where did it come from?

What are its limitations and biases?
Who is using it for what?
How has it changed over time?

Theater documentation exists to satisfy audit requirements or create the appearance of governance without actually helping anyone make better decisions about data use.

A retail company built a data catalog that was genuinely useful. Each data set had business descriptions in plain language, technical schema, source system and refresh frequency, known quality issues and limitations, contact person who could answer questions, examples of current uses and access procedures. This took a few hours per data set to create and required minimal maintenance.

People used this catalog because it answered real questions. "What customer data can I access?" "Is this sales data complete or missing some channels?" "Who else is working with product hierarchy data that I could learn from?"

They had previously had comprehensive data governance documentation that nobody read because it was written to satisfy compliance requirements rather than serve user needs. The new catalog took less effort to maintain and created more value.

POLICY THAT EVOLVES WITH USE

Data governance policies should not be static documents created once and enforced forever. They should evolve based on how people work and what problems emerge.

Start with minimal necessary policies: how to manage sensitive data, who can approve what, what's prohibited for legal or ethical reasons. Add policies when you discover gaps or when new situations create risks that are not addressed.

Remove or simplify policies that create friction without reducing risk. If nobody can explain why a policy exists or what risk it mitigates, question whether it is necessary.

An energy company reviewed their data policies quarterly. Each review looked at: What policies did people struggle to follow? What workarounds emerged? What risks materialized that policies did not prevent? What policies did people not even know existed?

They eliminated about thirty percent of their policies as unnecessary or outdated. They simplified others to be more practical. They added a few new policies addressing risks they had discovered. The governance program became lighter and more effective simultaneously.

PRIVACY, SECURITY, AND ETHICS SIMPLIFIED

These topics can spiral quickly into complexity. Legal requirements, technical controls, ethical frameworks, risk assessments. You need to address them seriously without getting paralyzed by complexity.

Privacy by design

Build privacy protection into how you design AI systems rather than adding it as an afterthought.

This means collecting only data you need for defined purposes, not everything you could collect. Store data with appropriate access controls from the start. De-identify or anonymize data when full identity is not necessary. Build systems that let people understand what data you are using and how. Create processes to delete data when it is no longer needed or when people request deletion.

A social media company building content recommendation AI faced privacy decisions early. They could use all user behavior data to optimize recommendations. They chose to limit data use to what users explicitly shared publicly and interactions users chose to make visible. More restrictive than necessary legally; aligned with their privacy principles and reduced risk.

This decision constrained some AI capabilities. It also meant they could explain clearly to users what data powered recommendations, could honor deletion requests simply, and had lower risk exposure to privacy regulation changes.

Security appropriate to sensitivity

Not all data needs the same security level. Match security controls sensitivity and risk.

Public information or data that is de-identified to the point where it reveals nothing sensitive can have basic security, enough to prevent tampering or unauthorized modification, not necessarily tight access control.

Personal information, confidential business data, or information subject to regulations needs stronger security: encryption, access logging, multi-factor authentication, regular security reviews.

Highly sensitive data, health records, financial details, proprietary IP, needs comprehensive security: strong encryption at rest and in transit, limited access, detailed audit trails, regular penetration testing.

Ethical frameworks that guide decisions

Ethics in AI is genuinely complex. You are making decisions about fairness, transparency, autonomy, accountability, and impact that do not have clear right answers.

You need frameworks that help you make these decisions consistently rather than ad-hoc, but frameworks are simple enough to use rather than gathering dust on shelves.

Start with basic principles: Are we treating people fairly, or are we creating differential outcomes based on characteristics that should not matter? Are we transparent about how AI systems work and what they are doing, or are we obscuring this? Are we giving people agency in how AI affects them, or are we making decisions about them without recourse? Are we accountable for harms AI systems might cause?

Then create lightweight processes to apply these principles: brief ethical review of new AI use cases, asking these questions explicitly. Diverse perspectives in the review, not just technologists, but people who understand business, regulatory, and social implications. Clear escalation paths when ethical concerns emerge. Regular retrospectives on deployed AI to check whether ethical principles are being held up in practice.

DATA READINESS ASSESSMENT

Before you commit resources to AI initiatives, honestly assess whether your data foundation will support them.

This assessment needs to be specific to planned use cases, not abstract evaluation of overall data maturity. Different AI applications need different data, and you might be ready for some use cases while not ready for others.

THE SIX-QUESTION DATA READINESS CHECK

For each AI use case you are considering, ask:

Does the data exist? It seems obvious, yet projects sometimes launch based on assumptions that needed data exists when it doesn't. Verify existence before you commit to building something.

Can we access it? Data might exist but be locked in systems you cannot easily extract from, owned by parts of the organization that will not share it, or controlled by third parties with restrictive terms.

Is there enough of it? Machine learning needs volume. How much volume depends on problem complexity and data quality. As rough guidance, simple problems might work with thousands of examples. Complex problems often need millions. Be honest about whether you have adequate volume.

Is quality sufficient? This is subjective, but you need realistic assessment. Look at sample data. How complete is it? How accurate? How consistent? Are there obvious problems that would prevent pattern learning?

Can we use it legally and ethically? Check privacy regulations, contractual restrictions, ethical guidelines. Do not assume because you possess data that you can use it for AI.

Can we maintain it? Initial data might be good, but can you maintain quality, handle changes, refresh data as needed, incorporate feedback from model performance?

An agricultural technology company ran this assessment for a crop yield prediction project. Data exists? Yes, years of yield data from customer farms. Can we access it? Mostly, though some customers had data in formats requiring manual conversion. Is there enough? Adequate volume for major crops, insufficient for specialty crops. Is quality sufficient? Responses were variable. Some customers had detailed data, others had minimal records. Can we use it legally? Yes, customers had granted rights. Can we maintain it? Yes, data flows were already established for other purposes.

Verdict: ready to proceed for major crops with data quality improvement work planned, not ready for specialty crops without substantially more data collection.

This kind of specific assessment prevents the unpleasant surprise of discovering mid-project that does not support what you are trying to build.

BUILDING DATA CAPABILITY INCREMENTALLY

You are not going to fix all data problems before starting AI. You also cannot ignore data problems and hope they will not matter. The practical path is building data capability incrementally, improving what you need for current AI initiatives while establishing patterns and infrastructure that support future work.

THE USE-CASE-DRIVEN BUILD SEQUENCE

Start with data infrastructure that supports your highest-priority AI use case. Build it properly. Not just a one-off solution but an approach you can replicate. Document what you did and why. Establish standards for how this type of data should be managed.

When you tackle your second use case, reuse what you built for the first wherever applicable. Add new infrastructure only where the second use case has different requirements. Again, document and standardize.

By the third and fourth use cases, patterns emerge. You have built reusable components, established standards, developed expertise. New use cases get easier because you are building a foundation rather than starting from nothing each time.

OPPORTUNISTIC DATA QUALITY IMPROVEMENT

Do not try to fix all data quality problems. Fix problems that matter for current AI use cases and fix them in ways that improve data rather than just for one project.

If you discover customer address data is inconsistent, and addresses matter for your current AI project, fix the address data. While you are at it, improve the data entry process so future addresses are better. Fix the problem once in a way that prevents recurrence.

If you find that product categorization is messy, and categorization matters for your current work, clean it up and establish better categorization maintenance. Other projects benefit from this improvement even if product categories were not their focus.

A healthcare system discovered date inconsistencies while building an AI application for appointment no-show prediction. Appointment dates were recorded inconsistently across different scheduling systems. They could have cleaned just the appointment data. Instead, they fixed the underlying system integration issues causing date inconsistencies. This improvement helped the no-show prediction project and fixed problems affecting appointment reminders, billing, and reporting.

Opportunistic improvement compounds. Each AI project leaves data infrastructure slightly better than it found it. Over time, data quality and consistency improve substantially without requiring a separate data quality program.

RESISTING THE PERFECT INFRASTRUCTURE TRAP

Organizations sometimes decide they need to build perfect data infrastructure before they can do real AI work. Enterprise data warehouse, master data management, comprehensive data governance, data quality tools, metadata management, data catalogs.

These can all be valuable eventually. Requiring them before starting AI work means you will spend years building infrastructure without delivering AI value. By the time infrastructure is ready, business priorities will have changed, the team that understood the original use cases will have moved on, and organizational patience with AI will be exhausted.

Build minimum viable infrastructure, deliver value, improve infrastructure incrementally based on real needs rather than theoretical completeness.

COMMON DATA PITFALLS

Even organizations that take data seriously make predictable mistakes. Knowing these patterns helps you avoid them.

Underestimating data preparation effort

Machine learning courses make it look easy. Clean data set, plug it into an algorithm, get predictions. Reality is messier. Data preparation typically consumes sixty to eighty percent of AI project effort.

You are extracting data from systems that were not designed for export. You are resolving inconsistencies, overseeing missing values, merging data from various sources with different formats and timing. You're engineering features that might be predictive. You're creating labeled training data. This all takes time.

Plan for it. Budget for it. Do not assume data preparation is a quick preliminary step before the real AI work. It is real AI work.

Ignoring data drift

Data changes over time. Customer behavior shifts. Products evolve. Market conditions fluctuate. Business processes are getting updated. Your data distributions drift, and models trained on historical data become less accurate on current data.

Organizations sometimes deploy models and assume they will keep working indefinitely. They will not. You need monitoring to detect drift and processes to retrain or adapt models when drift becomes significant.

Confusing correlation with causation in data

Machine learning finds patterns. It does not distinguish between patterns that represent genuine causal relationships and patterns that are coincidental correlations. Your job is making that distinction.

A retailer found their model predicted high customer lifetime value based partly on whether customers had purchased a specific product category. They almost designed a promotion around this category to acquire high-value customers. Closer investigation revealed the category was baby products. The real pattern: customers in a certain life stage had high lifetime value, and they happened to buy baby products because they had babies. Promoting baby products to people without babies would not create high-value customers.

Always ask: does this pattern make business sense? Could it be spurious correlation rather than meaningful relationship? What would happen if we acted on this pattern?

Insufficient data security and access control

Data breaches are expensive. Lost customer trust, regulatory penalties, competitive disadvantage if proprietary data leaks. Yet organizations sometimes build AI systems without adequate data security because security seems like a technical implementation detail, not a strategic priority.

It is a strategic priority. Build security from the start. It is much harder to add later, and the cost of getting it wrong can be catastrophic.

Making Data Foundation Real

The successful twenty percent treat data foundation as strategic investment, not technical overhead. They invest early, improve continuously, and make data quality everyone's responsibility rather than just IT's problem.

They are honest about data readiness. If data does not support a use case, they either fix the data first or choose a different use case rather than pretending data problems will not matter.

They build incrementally based on real AI needs rather than trying to create perfect data infrastructure before proving AI value.

They establish just enough governance to prevent problems without creating bureaucracy that slows everything down.

Most fundamentally, they recognize that AI without good data is just expensive randomness. The algorithm might be brilliant. If data is poor, results will be poor. No amount of technical sophistication compensates for bad data foundation.

Your AI transformation does not start with model architecture or platform selection or vendor evaluation. It starts with honest assessment of whether your data will support what you are trying to build, and if not, do the unglamorous work of fixing that foundation before you build on top of it.

The organizations that skip this work, or half-do it, or convince themselves that somehow their project will be the exception where data problems do not matter. They are the ones contributing to the eighty percent failure rate.

The ones that succeed are the ones who accepted that AI transformation requires data transformation, and they did the arduous work of building proper foundations before they tried to build AI solutions.

Not exciting work. Critical work. The difference between AI that creates value and AI that creates expensive disappointment.

CHAPTER 11

MEASURING SUCCESS AND ROI

Six months into their AI transformation, the CFO could not shake the feeling that something was off. The project team kept reporting "progress." The technology vendor assured them everything was "on track." Yet when she asked the question that mattered most: "What's this actually worth to us?", the room went quiet.

Were they building something valuable, or were they just building something?

This is the question that haunts every AI initiative, usually far too late in the game. Companies pour resources into transformation efforts, celebrate technical milestones, and highlight proof-of-concepts at board meetings. Then, somewhere around month nine or twelve, someone finally asks about return on investment. By then, the project has its own momentum, its own political constituency, its own reasons for existing that have nothing to do with business value.

The uncomfortable truth? Most organizations are measuring the wrong things entirely.

THE VANITY METRICS TRAP

Walk into any AI project review, and you'll hear numbers that sound impressive. "We've deployed 47 machine learning models." "Our data scientists have processed twelve million records." "We've reduced processing time by 43%." These metrics share a common flaw: They measure activity, not value.

Activity is seductive because it is easy to measure and consistently positive. Models deployed always go up. Data processed always increases. These numbers make good slides. They suggest progress, momentum, forward motion. Executives who do not know what else to measure latch onto them gratefully.

A manufacturing company spent eighteen months building what they called an "AI-powered quality control system." Every monthly review highlighted impressive metrics: images analyzed, defects detected, model accuracy improvements. The numbers climbed steadily upward. Only problem? The defect rate in shipped products had not budged. The AI was identifying defects that human inspectors were already catching. It was solving a problem the company did not actually have.

When vanity metrics dominate, organizations optimize for the wrong outcomes. Data science teams focus on technical sophistication rather than business impact. Project managers emphasize completion percentages overvalue delivery. Everyone looks busy, but no one can explain what all this motion is worth.

The shift from vanity to value requires asking different questions. Not "How many models did we build?" but "How much cost did we eliminate?" Not "How much data did we process?" but "What decisions improved?" Not "How accurate is the algorithm?" but "How much more money are we making?"

These questions are harder to answer. They require connecting technical outputs to business outcomes, which means understanding causation, not just correlation. They demand that someone trace the path from AI capability to customer behavior to budgetary impact. This work is unglamorous, often tedious, and absolutely essential.

THE FOUR DIMENSIONS OF AI VALUE

AI creates value in ways that do not always show up in traditional financial metrics, at least not immediately. A complete measurement system tracks four distinct dimensions: financial impact, operational improvement, strategic positioning, and organizational capability.

Financial impact is the easiest to understand and the hardest to measure precisely. This includes direct revenue increases, cost reductions, capital efficiency improvements, and risk mitigation. A retail bank's fraud detection system does not just catch fraudsters; it reduces losses, lowers investigation costs, improves customer trust, and decreases regulatory penalties. Each of these translates to dollars, though the translation requires different methods and involves various levels of certainty.

Revenue attribution is particularly tricky. When a recommendation engine suggests products and customers buy them, how much credit does the AI deserve? The customer might have purchased it anyway. The product might have been promoted elsewhere. The brand itself carries weight. Clean attribution is often impossible, but reasonable approximation is typically good enough. The question is not whether the AI deserves 100% credit; it's whether the incremental lift justifies the investment.

Operational improvement measures efficiency, speed, quality, and capacity gains. These metrics sit closer to the actual AI implementation, making them easier to measure but sometimes harder to value. A legal firm's contract review AI processes documents 60% faster than human lawyers. That is operational improvement. The value depends on what the firm does with that speed: serve more clients, reduce turnaround time, reallocate lawyers to higher-value work, or simply maintain current capacity with fewer people.

The mistake many companies make is measuring operational improvement but not completing the chain to financial value. Faster processing means nothing if the time saved vanishes into organizational slack. Improved accuracy matters only when errors currently cost money or opportunity. Higher capacity is worthless if demand does not exist to fill it.

Strategic positioning is about competitive advantage and future optionality. An insurance company builds AI capabilities not just for immediate returns but to compete in a market where data-driven pricing and underwriting become table stakes. A logistics company invests in route optimization AI because the industry is moving toward autonomous vehicles, and they need the foundation now. These investments may not pay off for years, but they are building capabilities that the organization will need to remain viable.

Measuring strategic value requires different thinking. The question is not "What's the ROI this quarter?" but "What happens if we don't do this?" Sometimes the answer is "nothing immediate." Sometimes it's "we lose the ability to compete in three years." Both are valid metrics, but only one shows up in traditional financial analysis.

Organizational capability tracks learning, culture change, and institutional knowledge. The first AI project teaches lessons worth far more than its direct outputs. Teams learn to work with data differently. Managers develop intuition for what AI can and cannot do. The organization builds confidence in its ability to deploy modern technology. These capabilities compound over time, making subsequent AI initiatives faster, cheaper, and more likely to succeed.

A pharmaceutical company's first AI project, a modest tool for predicting drug interactions, delivered reasonable financial returns. Its real value lay elsewhere. The project created a cadre of people who understood AI implementation, established governance processes, built relationships with technology partners, and demonstrated to skeptical scientists that AI could actually work. When the company later tackled more ambitious AI initiatives, they moved twice as fast and failed half as often. That acceleration, that organizational learning, is value that never appears in a project ROI calculation but matters immensely to enterprise success.

BUILDING YOUR MEASUREMENT SYSTEM

Effective measurement starts before the first line of code is written. The moment an AI initiative is proposed, three questions must have clear answers: What are we trying to improve? How will we measure improvement? What level of improvement justifies the investment?

This sounds obvious, yet most projects skip this foundation entirely. Teams dive into technical specifications without defining success criteria. They build solutions to problems that were never quantified. They make assumptions about value that no one bothers to validate. Then, when someone eventually asks about ROI, they are stuck retrofitting metrics onto initiatives that weren't designed with measurement in mind.

The **Business-First Measurement Canvas** provides structure for this foundation work. It's a single page that forces clarity on the metrics that matter:

Current State: What is the baseline performance? How much does the problem cost today? What's the current process efficiency? Establish these numbers with precision because they're the reference point for everything that follows. Many projects fail this basic step, comparing their results to vague estimates or assumed baselines that turn out to be wrong.

Target State: What is the goal? A specific, measurable improvement target with a period. Not "reduce costs" but "reduce customer service costs by 15% within 12 months." Not "improve quality" but "decrease defect rate from 3.2% to 2.1% by Q4." Specificity forces clarity. If you cannot define the target precisely, you do not understand the problem well enough to solve it.

Leading Indicators: What early signals suggest you are on track? These are metrics you can measure weekly or monthly that predict eventual success. For a customer churn prediction model, leading indicators might include model accuracy, percentage of predictions reviewed by account managers, and conversion rate of intervention attempts. These do not directly measure business value, but they indicate whether the system is working as designed.

Lagging Indicators: What is the ultimate business outcome? These are the metrics that matter to P&L: revenue, costs, margin, customer retention, market share. They might take quarters to measure definitively, but they are why the project exists. Leading indicators help you steer; lagging indicators tell you whether you arrived.

Measurement Frequency: How often will you check? Daily dashboards for operational metrics, monthly reviews for business impact, quarterly deep dives for strategic value. Different metrics need different rhythms. Over-measuring wastes time; under-measuring means flying blind.

Ownership: Who is responsible for each metric? Not just collecting the data but explaining what it means and acting when numbers move wrong. Measurement without ownership is theater.

One crucial discipline: separating correlation from causation. The AI initiative and business improvement might coincide without one causing the other. Markets shift. Competitors stumble. Other organizational changes take effect. The AI gets credit (or blame) for results it did not create.

Isolating AI impact requires control groups, or careful before-after analysis that accounts for other variables. A retail chain rolling out inventory optimization AI should measure stores with the system against comparable stores without it. A bank assessing a new credit scoring model should compare approved customers against a held-out control group. These methods are not always possible, but when they are, they are the difference between knowing and guessing.

THE AI ROI CALCULATOR

Return on investment sounds straightforward: divide benefits by costs, get a percentage. The reality is messier because both numerator and denominator hide complexity.

The actual cost of AI includes technology expenses, yes, but also data preparation, model development, integration work, change management, training, ongoing maintenance, and the opportunity cost of the team's time. Most organizations dramatically underestimate these costs because they focus only on the most visible line items.

A financial services firm budgeted $800,000 for an AI project: $500,000 for software and consulting, $300,000 for hardware infrastructure. The actual cost? $2.1 million. They'd forgotten to include the cost of six months of their data engineering team's time, the business analysts mapping requirements, the legal review of algorithmic bias, the customer service training on AI-assisted workflows, and the senior executive time spent managing vendor relationships and internal politics. These costs were real, even if they did not appear in the project budget.

Benefits are even trickier to quantify. Some are direct and measurable: reduced labor costs, increased sales, lower error rates. Others are indirect: faster decision-making, improved customer satisfaction, better risk management. The indirect benefits often matter more than the direct ones, but they are harder to value with precision.

The **AI ROI Calculator** breaks down both sides of the equation systematically:

Cost Components

One-Time Costs:

- Technology acquisition and licensing
- Integration and development
- Data preparation and infrastructure
- Training and change management
- Process redesign
- Initial consulting and support

Recurring Costs:

- Ongoing licensing and subscription fees
- Infrastructure and computing resources
- Maintenance and updates
- Support and troubleshooting
- Model retraining and improvement
- Governance and monitoring

Benefit Components

Direct Financial Benefits:

- Revenue increase from new capabilities
- Cost reduction from automation
- Efficiency gains from faster processing
- Quality improvements reducing waste or rework
- Risk reduction avoiding losses

Indirect Benefits:

- Improved decision quality
- Enhanced customer experience
- Faster time-to-market
- Better resource allocation
- Increased organizational learning
- Competitive positioning

Intangible Benefits:

- Brand perception improvement
- Employee satisfaction
- Innovation culture development

- Market credibility
- Strategic optionality

For each benefit, estimate the impact range: pessimistic, realistic, optimistic. This acknowledges uncertainty rather than pretending you know the precise value. A customer service chatbot might reduce call volume by 15-35%. The realistic case is 25%, but showing the range helps stakeholders understand the bet they are making.

Calculate ROI across multiple periods: Year 1, Year 3, Year 5. AI investments often look terrible in Year 1 (excessive costs, modest returns) and excellent by Year 3 (costs stabilized, benefits scaled). Organizations that evaluate only short-term returns systematically underinvest in capabilities that take time to mature.

One metric deserves special attention: payback period. How long until cumulative benefits exceed cumulative costs? For most AI projects, the answer is 18-36 months. Projects promising six-month payback are either incredibly simple implementations or wildly optimistic projections. Projects showing five-year payback face too much uncertainty to justify. The sweet spot for most AI investments sits somewhere in the middle: long enough to build something meaningful, short enough to maintain confidence.

DASHBOARDS THAT DRIVE ACTION

Data without decisions is just noise. The purpose of measurement is not to generate reports; it is to enable better choices. An effective AI dashboard surfaces the information that demands attention and makes clear what action that information suggests.

Most dashboards fail by including too much. Fifty metrics measure every conceivable dimension of system performance. Someone worked hard to collect all that data, so it gets displayed. The result? Nobody looks at any of them. When everything is highlighted, nothing is.

The right dashboard includes five to seven key metrics, no more. These metrics should answer three questions: Is the system working technically? Is it delivering business value? Are there problems requiring intervention?

A logistics company's route optimization AI dashboard tracked just six numbers:

- Percentage of routes successfully optimized (technical health)
- Average delivery time versus baseline (operational impact)
- Fuel cost per delivery versus baseline (monetary impact)
- Driver satisfaction score with AI recommendations (adoption indicator)
- Manual override rate (trust indicator)
- Cost per optimized route (efficiency metric)

Each metric had green, yellow, red thresholds. Green meant "keep going." Yellow meant "pay attention." Red meant "take action now." No interpretation required. No analysis paralysis. The dashboard told them whether the AI was working and whether it was creating value.

Good dashboards also show trends, not just snapshots. A single week's numbers mean little. The pattern over time reveals whether things are improving, degrading, or staying flat. An upward trend in manual override rate, even if the absolute number seems acceptable, signals growing mistrust that will eventually kill the project. Catching it early creates opportunity for intervention.

Context matters too. A 12% reduction in processing time sounds modest until you know the industry average is 5%. A 2% improvement in prediction accuracy seems tiny until you realize it translates to $8 million in prevented losses. Dashboards should provide this context explicitly and not assume viewers will know it.

The best dashboards include qualitative insights alongside quantitative metrics. A single customer quote explaining why the AI recommendation was helpful. A brief summary of the week's most interesting edge case. A highlight of an unexpected use case that emerged organically. These human stories make the numbers meaningful and keep teams connected to the actual value they are creating.

WHEN TO SCALE, WHEN TO STOP

Measurement exists to inform decisions, and no decision matters more than whether to scale an AI initiative. Organizations waste vast resources either scaling things that do not work or failing to scale things that do.

The **Scale Decision Framework** creates structure around this choice:

Scale signals indicate an initiative is ready for broader deployment:

- Consistent business impact meeting or exceeding targets across multiple measurement periods
- Technical stability with error rates below thresholds and minimal manual intervention required
- User adoption without coercion, with organic spread and positive feedback
- Clear path to profitability with benefits accelerating as costs stabilize
- Organizational readiness with processes, skills, and governance in place

These signals should appear together, not in isolation. A technically perfect system that users hate will not scale successfully. An enthusiastically adopted tool that loses money is not sustainable. All dimensions need to align.

Stop signals suggest an initiative should be terminated or fundamentally rethought:

- Persistent failure to meet minimum success criteria despite multiple adjustment attempts

- Fundamental technical limitations that prevent the system from working as designed
- User resistance that does not diminish despite training and support
- Cost trajectory that makes profitability implausible even at scale
- Misalignment with strategic direction or changing business priorities

Stopping is not failure; it's intelligent resource allocation. The ability to kill projects that are not working is what enables organizations to invest more in projects that are. Yet most companies struggle to stop AI initiatives once they have started, especially if they've already invested significant resources.

The sunk cost fallacy runs rampant in AI. "We've already spent $3 million; we can't stop now." Yes, you can. The $3 million is gone regardless. The only question is whether spending another $3 million creates value. If not, stop. The first $3 million was the price of learning what does not work. That is valuable information, but only if you act on it.

A pharmaceutical company spent two years building an AI system to predict optimal drug dosing. The science was sound, the data was good, the models were accurate. The problem? Doctors would not use it. Despite training, incentives, and leadership mandates, adoption never exceeded 12%. After multiple change management interventions failed to move the needle, the company finally acknowledged reality: the system was solving a problem, doctors did not think they had. They stopped the initiative, redeployed the team to a different AI project, and counted the experience as tuition paid for organizational learning.

Between "scale" and "stop" lies "pivot", adjusting the approach while maintaining the core initiative. Maybe the original use case was too ambitious, but a simplified version would work. Maybe the technology was right, but the implementation approach was wrong. Maybe the business case exists in a different department than originally planned. Pivoting requires the same honest assessment as stopping, but with the belief that a modified approach can succeed.

The decision between scale, pivot, or stop should be made quarterly for new initiatives, annually for mature ones. Scheduled decision points prevent both premature abandonment and indefinite limbo. Everyone knows when the evaluation is coming, what criteria will be used, and who makes the call.

CONTINUOUS IMPROVEMENT THROUGH MEASUREMENT

AI systems do not stay effective through benign neglect. Model accuracy degrades as the world changes. Users need to evolve. Competitors respond. What worked brilliantly last year delivers mediocre results today. Sustained success requires continuous measurement and adjustment.

The improvement cycle runs on a regular rhythm: measure, analyze, act, measure again. A retail pricing algorithm might run this cycle weekly. An insurance underwriting model might operate quarterly. The frequency depends on how quickly the environment changes and how much data you generate, but the cycle itself never stops.

Measure: Collect current performance data across all key metrics. Not just automated system logs, but user feedback, business outcomes, competitive context.

Analyze: Compare current performance to baselines and targets. Identify degradation patterns. Surface anomalies. Investigate why numbers moved, not just that they moved. A drop in model accuracy might reflect data quality issues, conceptual drift in underlying patterns, or system problems corrupting inputs. Each cause requires a different response.

Act: Implement improvements based on analysis. This might mean retraining models with fresh data, adjusting business rules, fixing data pipelines, or changing how users interact with the system. Track what you change so you can measure impact.

Measure Again: Verify that actions had intended effects. Sometimes improvements help. Sometimes they hurt. Sometimes they do nothing at all. You need measurement to distinguish between these outcomes.

An insurance company's fraud detection AI initially achieved 89% accuracy. Over six months, performance drifted down to 76%. Investigation revealed that fraud patterns had evolved; the model was catching yesterday's schemes while missing today's. Retraining with recent data boosted accuracy to 91%, which was better than the original baseline. Without continuous measurement, the drift would have continued unnoticed, and the company would have been detecting an increasingly obsolete form of fraud.

This cycle of measurement and improvement creates organizational muscle memory. Teams develop intuition for what normal performance looks like, what abnormal patterns mean, and how to respond quickly.

THE ROI CONVERSATION

Numbers do not speak for themselves. Even the most compelling ROI analysis requires human interpretation, context, and storytelling. The measurement system generates data; your job is to transform that data into understanding.

Different audiences need different conversations. The CFO cares about payback period and present net value. The COO wants operational efficiency metrics and capacity impacts. The CEO is thinking about strategic positioning and competitive advantages. Frontline managers need to know how their work contributes and what behaviors to encourage. One ROI story does not fit all.

The structure of an effective ROI presentation follows a narrative arc:

Problem: What business challenge did this AI initiative address? Why did it matter? What was at stake?

Solution: What did we build? How does it work, explained simply? What makes this approach different or better?

Results: What happened? Show the metrics that matter to this audience. Explain magnitude and compare them to targets. Acknowledge what worked and what did not.

Insights: What did we learn? What surprised us? What would we do differently next time? This is where organizational learning becomes explicit.

Implications: What should we do now? Scale? Pivot? Apply lessons elsewhere? The measurement is not the point; the decision is.

Honest reporting includes both successes and failures. Every AI initiative encounters problems. Pretending otherwise destroys credibility. The executive team wants to know if you understand what went wrong and what you are doing about it. Leaders trust people who acknowledge reality more than people who spin everything positively.

This kind of straight talk requires organizational culture that does not punish messengers. If people fear reporting unwelcome news, they will hide problems until they become disasters. If leaders treat every setback as a failure of character, teams will either inflate successes or avoid ambitious goals. Healthy measurement cultures celebrate learning, even when the lesson is "this didn't work."

MAKING MEASUREMENT SUSTAINABLE

The measurement system itself needs maintenance. What starts as a crisp, focused dashboard gradually accumulates metrics until it becomes cluttered and useless. Data pipelines break. Thresholds that made sense at launch become obsolete. Manual data collection that was manageable at pilot scale becomes impossible at enterprise scale.

Sustainable measurement requires three disciplines:

Simplicity: Regularly prune metrics that no longer drive decisions. If no one acts on a metric, stop measuring it. New metrics can be added, but only if others are retired. The dashboard stays lean.

Automation: Eliminate manual data collection wherever possible. If someone is spending hours each week pulling numbers into spreadsheets, automate it or stop measuring it. Measurement overhead should be roughly 5% of project effort, not 30%.

Evolution: Revisit measurement strategy annually. Are we measuring the right things? Do these metrics still align with business goals? Have we learned enough to simplify? What new questions do we need answers to? The measurement system should improve along with everything else.

A global manufacturer started with seventeen metrics tracking their warehouse automation AI. After six months, they realized they looked at only four regularly and acted on only two. They simplified to five metrics total: picking accuracy, throughput per hour, system uptime, manual override rate, and cost per pick. The dashboard became clearer, faster to review, and more useful. Team meetings focused on discussing implications rather than explaining metrics.

The measure of any measurement system is whether it improves decisions. If you are generating dashboards that nobody uses, tracking metrics that do not matter, and producing ROI analyses that gather dust, you're measuring wrong. The goal is not to measure more; it's to measure what matters and act on what you learn.

AI succeeds when it delivers real value, not when it deploys sophisticated technology. Knowing the difference requires measurement that is honest, relevant, and actionable. This chapter provided frameworks for building that measurement capability. Now the question becomes: What are you going to measure, and what will you do with what you discover?

The next chapter addresses the ultimate measure of AI success: building an organization that can continue learning, adapting, and thriving as technology evolves and markets shift. Because the real value isn't any single AI implementation. It is becoming the kind of organization that gets better over time.

CHAPTER 12

THE FUTURE-PROOF ORGANIZATION

A European insurance company had spent eighteen months building their AI capability. They had hired data scientists, bought platforms, and trained models. Then GPT-4 dropped, and half of what they had built became not quite obsolete but suddenly there was a better way to do things that they had not even considered. In this case it wasn't technology that aged poorly, it was their thinking.

That is the knife edge every organization walks now. You cannot ignore AI and hope to compete. Yet you also can't build your entire strategy around specific tools that might be superseded within eighteen months. The question isn't just how to implement AI successfully today. It's how to build an organization that can keep succeeding as the technology morphs into something unrecognizable.

The companies that thrive over the next decade will not necessarily be the ones with the most sophisticated AI implementations right now. They'll be the ones that learned how to learn, that built the organizational muscles for continuous adaptation, which figured out how to absorb technological change without collapsing into chaos every time something shifts.

This is harder than it sounds. Most organizations are built for stability, not transformation. Processes ossify and hierarchies protect themselves. Success breeds attachment to whatever produces that success. The very capabilities that made you excellent at the old game often prevent you from playing the new one well.

BEYOND SOLUTIONS, TOWARD CAPABILITY

There is a pattern I've watched repeat itself across industries. A company identifies an AI opportunity, builds a solution, deploys it successfully, and celebrates. The solution works beautifully for two years, sometimes three. Then market conditions shift, or the technology leaps forward, or competitors do something unexpected. The solution that was perfect becomes merely adequate, then inadequate, then a liability. The company needs to be rebuilt, but the team that built the original solution has moved on, the vendors have changed, the internal knowledge has evaporated.

They built a solution. They needed to build capability.

Solutions are specific. They solve particular problems with particular tools at particular moments. Capability is general. It is the organizational ability to identify problems, evaluate options, implement solutions, and do it again when circumstances change. Solutions depreciate. Capability compounds.

A telecommunications company in Southeast Asia learned this distinction in an expensive way. They spent substantial resources building a customer churn prediction model. State of the art for its time, genuinely impressive technical work. It predicted which customers were likely to cancel service with reasonable accuracy, allowing the retention team to intervene. The model delivered solid ROI for about eighteen months.

Then the market changed. New competitors entered with different pricing models. Customer behavior has shifted. The factors that predicted churn in 2022 bore little resemblance to what mattered in 2024. The model's accuracy degraded steadily. Worse, the team that had built it had been disbanded after launch, their skills reassigned to other priorities. The company had the solution, but they had lost the capability to evolve it.

Compare that to a retail bank that took a different approach. They built a fraud detection system, sure, but they invested equally in building the internal capability to understand, maintain, and improve that system over time. They kept the core team together. They documented not just what they built but why they made each decision. They established regular cycles for model retraining and evaluation. They created clear ownership for ongoing improvement.

When fraudsters shifted tactics, as fraudsters always do, the bank adapted within weeks. When new data sources became available, they integrated them quickly. When better algorithms emerged, they evaluated and adopted them systematically. Five years later, their fraud detection is unrecognizable from the original implementation, but it's never stopped working. They did not build a solution that aged out. They built a capability that keeps getting better.

The shift from solution-thinking to capability-thinking requires different mental models. You stop asking "What AI tool do we need?" and start asking "What organizational muscles must we develop?" You measure success not just by what the AI accomplishes today but by whether your organization is more capable of AI deployment tomorrow than it was yesterday.

This shows up in budget decisions. Solution-thinking puts most money into initial development, then minimal maintenance. Capability-thinking invests continuously in the platform, the team, the knowledge base, the processes. It costs more upfront, looks less efficient on paper, and pays for itself ten times over when you need to adapt.

It shows up in hiring. Solution-thinking hires for specific projects. Capability-thinking hires for organizational learning. The question is not just "Can this person build the model we need?" It's "Can this person teach others how to think about these problems?"

It shows up in project planning. Solution-thinking has a clear endpoint: deployment. Capability-thinking treats deployment as a midpoint in an ongoing journey of improvement and adaptation.

THE ADAPTIVE ORGANIZATION MODEL

Organizations that stay relevant as AI evolves share certain structural characteristics. They are not all structured identically, but they've each found ways to embed adaptability into how they operate.

Distributed expertise rather than centralized control. Early in their AI journey, most companies concentrate all AI capability in a single team or department. This makes sense initially. You need to build critical mass, establish standards, create efficiency. Over time, though, centralization becomes a bottleneck. Every AI opportunity must flow through one overloaded team. Local knowledge in business units cannot combine with technical capability in the center.

Mature AI organizations push expertise outward. They maintain a center of excellence that sets standards, provides platforms, and solves common problems. Then they seed AI-literate people throughout the business who can identify opportunities, implement solutions, and collaborate with the center when needed. The center enables rather than controls.

A manufacturing conglomerate moved from five data scientists in headquarters to fifty AI-capable people across divisions over three years. Not all fifty are technical specialists. Some are business analysts who learned enough about AI to identify beneficial use cases. Some are process engineers who understand how to integrate algorithmic decision-making into existing workflows. Some are change managers who know how to help teams adopt AI-augmented processes. Together, they have deployed three times as many successful AI solutions as the original centralized team, because they are closer to the problems and can move faster.

Experimentation infrastructure that makes trying new things cheap and fast. The future is uncertain. Some AI capabilities will matter enormously; others will prove irrelevant. You cannot know in advance which is which. Organizations that oversee uncertainty well create space for controlled experimentation.

This means technical infrastructure were spinning up a new project takes hours, not months. It means budget processes that allocate some resources to exploration without requiring bulletproof business cases. It means cultural permission to try things that might not work. Most experiments fail. That is fine, as long as they fail quickly and cheaply, teaching you something useful.

A pharmaceutical company established what they called "innovation sprints." Any team could propose a two-week experiment testing whether AI could improve some aspects of their work. If the idea passed a basic sanity check, they got access to data, computing resources, and technical support. No formal ROI calculation required. No months of planning. Just try it and see.

Most experiments produced nothing useful. A few generated modest improvements. Two unlocked breakthrough capabilities that the company never would have pursued through formal planning processes. The total cost of the program? Less than a single failed traditional AI project. The learning value? Immeasurable. Dozens of employees across the organization now understand AI capabilities from firsthand experience rather than abstract training.

Structured learning from both successes and failures. Experience teaches only if you extract the lessons. Most organizations are terrible at this. They repeat the same mistakes across different projects because no one systematically captured what went wrong last time. They fail to replicate successes because the knowledge stays trapped in one team's heads.

Future-proof organizations treat learning as a deliberate process, not a happy accident. After every significant project, regardless of outcome, someone sits down and documents what happened, what worked, what did not, and what they would do differently. These lessons get shared, discussed, and integrated into how the next project runs.

This sounds tedious, and frankly, it often is. Nobody loves writing post-mortems. The payoff comes later, when a new team faces a similar challenge and can learn from five previous attempts instead of starting from nothing. Organizational memory becomes a strategic asset.

Portfolio thinking about AI investments. Adaptable organizations do not bet everything on one approach. They maintain a portfolio of AI initiatives at different risk levels, different time horizons, different degrees of novelty.

Some projects are conservative: proven technology applied to well-understood problems. Minimal risk, moderate return, high confidence. These pay the bills and maintain momentum.

Some projects are aggressive: experimental technology pursuing ambitious goals. Considerable risk, potentially transformational return, low confidence. Most will fail, but the ones that work might reshape the business.

Most projects sit somewhere in the middle: reasonable risk, solid potential, decent odds of success.

The portfolio approach prevents both timidity and recklessness. You are always learning, always having some successes, always pushing boundaries, but never betting the company on unproven technology. As circumstances change, you adjust the portfolio's composition. In stable times, shift toward more experimental work. When uncertainty rises, pull back to safer investments.

An energy company maintains roughly 60% of AI investment in incremental improvements, 30% in significant innovations, and 10% in moonshots. The percentages are not magic; they are what fit this company's risk tolerance and strategic position. The discipline is what matters: explicitly deciding how to balance exploration and exploitation rather than lurching between them reactively.

PREPARING FOR WHAT IS COMING

The AI landscape in five years will look nothing like today. You do not know exactly how it will change, and neither do I, despite what consultants might claim. What you can do is develop an informed perspective on where things are heading and position yourself to capitalize on those shifts rather than be blindsided by them.

Some trends look relatively certain, barring major surprises. AI will keep getting more capable across more domains. The cost of computational power will continue falling. The boundary between "human work" and "AI work" will keep shifting. These are not controversial predictions; they are extrapolations of consistent patterns.

Other developments are probable but less certain. Regulations will tighten, though the specifics vary by jurisdiction. Concerns about AI safety, bias, and societal impact will intensify. The competitive advantage from simply having AI will diminish as adoption becomes universal; the advantage will come from having better AI or using it more effectively.

Some possibilities are speculative but worth thinking about. Artificial general intelligence that matches or exceeds human capability across all domains. Fundamental breakthroughs in how we train and deploy models. Catastrophic failures that reshape public trust and regulatory approaches. Black swan events that nobody saw coming.

You cannot plan for every scenario. You can build an organization that manages uncertainty well. That means avoiding brittle strategies that work brilliantly under one set of assumptions but collapse if those assumptions prove wrong. It means maintaining optionality, keeping doors open, not locking yourself into irreversible commitments when you can avoid it.

The finance industry is watching quantum computing's evolution with particular interest. Quantum machines could potentially break current encryption methods, revolutionize risk modeling, and reshape algorithmic trading. Or technology might remain impractical for decades. Nobody knows.

Smart financial institutions aren't betting their strategy on quantum computing arriving on schedule. They are monitoring developments, building relationships with researchers, conducting small experiments to understand technology's potential, and ensuring their systems can adapt when quantum becomes relevant. They are ready to move fast if the technology matures, but they are not paralyzed waiting for it.

That is the posture you want across emerging technologies. Aware, prepared, flexible, but not distracted from delivering value today while you wait for tomorrow.

BUILDING AN INNOVATION CULTURE

The future-proof organization needs more than good processes and smart strategies. It needs a culture that embraces change rather than resists it. Culture is notoriously difficult to shift; it's the accumulated weight of thousands of individual decisions, behaviors, and assumptions. You cannot change it overnight. You can shape it deliberately over time.

Psychological safety is foundational. People will not experiment if failure means punishment. They won't share concerns if dissent is treated as disloyalty. They won't try innovative approaches if success requires following established playbooks. Innovation requires permission to challenge assumptions, including the assumptions of senior leaders.

This does not mean eliminating accountability or accepting sloppiness. It means distinguishing between failures that result from negligence or incompetence and failures that result from smart experiments that didn't pan out. The first type deserves consequences. The second type deserves analysis and learning.

A technology company instituted "failure forums" where teams presented projects that did not work, explaining what they tried, what happened, and what they learned. The forums weren't comfortable, exactly. Nobody enjoys admitting their project flopped. Yet over time, the practice normalized the idea that intelligent people sometimes pursue ideas that do not pan out, and that's okay as long as they learn from experience.

Curiosity as a professional expectation. The half-life of knowledge in AI is measured in months, not years. What you know today will be partially obsolete tomorrow. The only sustainable advantage is the ability to keep learning.

Organizations that navigate technological change well make learning part of the job, not something that happens on weekends or after hours. They allocate time for people to explore new tools, attend conferences, take courses, or just tinker with emerging technologies. They hire for learning capacity as much as current knowledge.

This requires genuine investment, not just lip service. "We value learning" means nothing if people are punished for missing deadlines because they took time to develop new skills. It means nothing if promotion depends entirely on execution of existing capabilities rather than growth of new ones.

Cross-functional collaboration breaks down the silos that prevent adaptation. AI transformation touches everything: technology, operations, strategy, legal, HR, finance. Organizations that treat it as purely a technology problem or purely a business problem struggle. Those that bring diverse perspectives together earlier and more often navigate change better.

A healthcare provider formed "innovation pods" mixing clinicians, technologists, administrators, and legal experts. Each pod had broad authority to identify opportunities and implement solutions within their domain. The pods were not always efficient. Doctors and data scientists sometimes spoke past each other. Getting consensus across disciplines took time. Yet the solutions they produced worked better and faced fewer implementation barriers because multiple perspectives shaped them from the start.

Comfort with ambiguity. The future is somewhat unknown. Plans will change. What seems like the right direction today might prove wrong tomorrow. People who need certainty before acting will freeze. People who can operate effectively despite uncertainty will thrive.

This is partly about hiring for tolerance of ambiguity. It is also about leadership modeling comfort with partial information and evolving strategies. When executives confidently present five-year detailed plans for AI, they're lying to themselves and everyone else. When they honestly acknowledge uncertainty while explaining their current best thinking and how they will adapt as circumstances change, they build credibility and resilience.

YOUR TWELVE-MONTH ACTION PLAN

Theory means nothing without execution. Here is what building a future-proof AI organization looks like in practice over the next year. This plan assumes you've already started your AI journey, perhaps implementing a few initial projects. If you haven't started yet, first work through the frameworks in earlier chapters. This chapter is about evolving from initial success toward sustained capability.

Months 1-3: Assessment and Foundation

Take honest inventory of where you are. Not where you wish you were or where presentations claim you are. Where you actually are.

Evaluate your current AI maturity using the framework from Chapter 5. Identify gaps between your current state and where you need to be. Prioritize the gaps that matter most: those that limit your ability to scale successful initiatives or adapt to changing conditions.

Assess your organizational structure for adaptability. How centralized is decision-making? How quickly can teams access resources for experimentation? How well do divergent functions collaborate? Where are the bottlenecks?

Review your recent AI initiatives, successful and failed. Extract lessons systematically. What patterns emerge? Which approaches consistently work? Which consistently fails? Who in your organization has developed real expertise?

Based on this assessment, define your capability-building priorities. Maybe you need to distribute expertise more widely. Maybe you need better experimentation infrastructure. Maybe you need to strengthen organizational learning. You cannot fix everything at once. Choose what matters most.

Document your current state honestly. This baseline matters. In twelve months, you will look back to see how far you've come.

Months 4-6: Capability Building

Start developing the organizational muscles you identified as priorities.

If you need distributed expertise, begin seeding AI-capable people across business units. This might mean training existing employees, rotating people from your central AI team into business roles or hiring new people with hybrid skills. Start small, learn what works, then scale.

If you need experimentation infrastructure, invest in platforms and processes that make trying new things faster and cheaper. Build templates, establish lightweight approval processes, allocate some budget to exploration. Run a few pilot experiments to validate that the infrastructure actually reduces friction.

If you need better learning mechanisms, start capturing lessons from projects systematically. Create simple templates for post-project reviews. Designate someone to compile insights into accessible formats. Share lessons openly, including failures.

If you need stronger culture, identify behaviors you want to encourage and make them visible. Celebrate smart experiments that failed. Recognize people who learned from mistakes. Promote individuals who develop others, not just those who deliver results.

These capability-building efforts will not show immediate ROI. They're investments in organizational muscle that pays off later. Expect resistance from people focused on short-term metrics. Persist anyway.

Months 7-9: Portfolio Optimization

Review your current portfolio of AI initiatives. Are you balancing risk appropriately? Do you have enough safe projects to maintain confidence? Enough ambitious projects to drive breakthrough potential? Are resources allocated thoughtfully or just historically?

Kill projects that aren't working. This is hard, politically fraught, and necessary. Resources and attention you free up enable better work elsewhere. Use the decision framework from Chapter 11 to evaluate each initiative honestly.

Scale projects that are working. Success deserves investment. Move winning initiatives from pilot to production, from one business unit to several, from regional to global. But scale thoughtfully. Forcing rapid expansion can break things that work well on a small scale.

Launch new experiments in areas you have been ignoring. If all your AI work focuses on efficiency, try something customer-facing. If everything is about automation, explore augmentation. Deliberately pursue diversity in your portfolio.

Months 10-12: Embedding and Acceleration

Make new capabilities and behaviors routine. What started as special initiatives should become normal ways of working.

Integrate capability-building into standard processes. New project charters include learning objectives. Budget reviews assess experimentation outcomes. Hiring criteria emphasize adaptability. These shouldn't be special programs anymore; they are just how things work here.

Measure your progress against the baseline from months 1-3. How has your AI maturity evolved? How much faster can you deploy new initiatives? How much more widely distributed is expertise? How much better are you at learning from experience? Celebrate gains, acknowledge gaps, adjust plans.

Look ahead to year two. What capabilities do you need next? What threats or opportunities are emerging? What will you focus on improving? The plan for next year should reflect what you learned this year.

THE TWENTY PRINCIPLES OF AI SUCCESS

We have covered enormous ground across these twelve chapters. Many frameworks, dozens of case studies, countless tactical recommendations. It is a lot to hold in your head. So let me distill everything down to twenty core principles. These are the ideas that, if you internalize nothing else, will dramatically improve your odds of AI success.

1. Start with problems, not technology. The best AI initiatives begin with clear business challenges that matter, not fascinating technology looking for application.

2. **Build capability, not just solutions.** Invest in your organization's ability to continuously deploy, improve, and adapt AI over time.

3. **Go slow to go fast.** Rushed implementations create technical debt and organizational resistance that slow everything down later.

4. **Measure what matters, not what is easy.** Focus on business value metrics, not technical achievement metrics.

5. **Treat data as a strategic asset.** Your data strategy determines your AI ceiling. Invest accordingly.

6. **Change management is not optional.** Technical excellence means nothing if people will not use what you build.

7. **Distribute expertise widely.** Centralized AI teams are necessary early but become bottlenecks. Push capability outward over time.

8. **Experiment constantly.** The future is uncertain. Try a lot of things. Most will fail. That is fine.

9. **Learn systematically.** Experience teaches only if you extract and share lessons deliberately.

10. **Balance your portfolio.** Mix safe bets, moderate risks, and moonshots according to your risk tolerance.

11. **Kill what does not work.** Stopping failed initiatives frees resources for successful ones.

12. **Scale what works.** Success deserves investment, but scale thoughtfully to avoid breaking what works.

13. **Simplify relentlessly.** Complexity is the enemy of adoption. Make things as simple as possible, but no simpler.

14. **Lead from the top.** Executive sponsorship means active engagement, not just budget approval.

15. **Build diverse teams.** Homogeneous groups produce blind spots. Mix disciplines, perspectives, and experiences.

16. **Maintain ethical guardrails.** AI capabilities outpace wisdom. Build values and oversight into everything.

17. Partner strategically. You cannot do everything yourself. Choose partners who share your values and complement your capabilities.

18. Invest in your people. Technology comes and goes. The ability to master recent technology persists. Upskill continuously.

19. Stay humble. AI is powerful but not magic. Acknowledge limitations honestly.

20. Adapt constantly. The only certainty is change. Build an organization that manages uncertainty well.

These principles are not complicated. None requires technical sophistication or massive resources. They are mostly about discipline, honesty, and sustained attention to what matters. Simple, not easy.

THE CHOICE BEFORE YOU

We are near the end now. You have read about why AI transformations fail and how successful one's work. You've been given frameworks, tools, templates, and case studies. You understand the principles that separate the 20% that succeed from the 80% that do not.

None of that matters if you do not act on it.

The path forward splits. One direction leads to becoming an AI-capable organization that continuously adapts and thrives. The other leads to becoming another cautionary tale, another statistic in someone's failure research, another company that thought AI would save them but never figured out how to make it work.

The difference between these paths is not luck. It is not funding, though resources help. It is not access to technology, though that matters. The difference is sustained commitment to doing the demanding work of organizational transformation. It is choosing reality over comfortable delusions. It's measuring honestly and acting on what you learn. It is building capability slowly when you want results quickly. It persists when nothing seems to be working and everyone's questioning whether this AI thing is worth the trouble.

Most organizations, if we are honest, will choose the easier path. They will launch AI initiatives with great fanfare. They'll make some initial progress, run into difficulties, get distracted by the next shiny thing, and drift back to familiar patterns. The AI transformation will join the strategic plan, the culture change program, and the digital initiative in the graveyard of good intentions that never quite happened.

Some organizations, though, a determined minority, will do the work. They will implement the frameworks imperfectly, learn from mistakes, adjust their approach, and keep going. Eighteen months from now, they will look nothing like traditional organizations in their industry. Three years from now, they'll have capabilities their competitors cannot match. Five years from now, they'll be defining what's possible while others struggle to catch up.

Which organization will yours be?

You now know what separates successful AI transformations from failed ones. You know the frameworks. You know the principles. You have seen the patterns. The question is not whether you have the knowledge. It is whether you'll use it.

Start small if you must. Pick one framework from this book and apply it seriously. Choose one principle and make it real in your organization. Take one action this week that moves you closer to becoming future proof. Then another next week. Then another after that.

Success compounds. So does failure. Every week you delay is a week of learning and adaptation you have lost. Your competitors are not waiting. Technology is not slowing down. The market is not getting more forgiving.

The blueprint is in your hands. The choice is yours.

Make it count.

SOURCES

IBM. (2023). *Global AI adoption index 2023.* IBM Institute for Business Value.

Informatica. (n.d.). The surprising reason most AI projects fail – and how to avoid it at your enterprise. Informatica.

Gartner. (2024). Gartner predicts that 30% of generative AI projects will be abandoned after proof of concept by end of 2025. Gartner Research.

Harvard Business Review. (2023). Why so many AI projects fail. Harvard Business Review.

McKinsey & Company. (2025). State of AI 2025: Why 88% adopt AI but only 6% transform their business. Lighthouse AI.

Fortune. (2024). MIT report: 95% of generative AI pilots at companies are failing. Fortune Magazine.

THE AI SUCCESS BLUEPRINT
– SCAN FOR FREE RESOURCES

www.ingramcontent.com/pod-product-compliance
Lightning Source LLC
LaVergne TN
LVHW022206200926
848481LV00006B/1862
9781049246697